Design and Debri

Design and Debris

A Chaotics of Postmodern American Fiction

Joseph M. Conte

The University of Alabama Press

Tuscaloosa and London

Typeface: Goudy

∞

The paper on which this book is printed meets the minimum requirements of American National Standard for Information Science–Permanence of Paper for Printed Library Materials, ANSI Z39.48–1984.

Library of Congress Cataloging-in-Publication Data

Conte, Joseph Mark, 1960–
Design and debris : a chaotics of postmodern American fiction / Joseph M. Conte.
p. cm.
Includes bibliographical references and index.
ISBN 0-8173-1114-9 (cloth : alk. paper) — ISBN 0-8173-1115-7 (pbk. : alk. paper)
1. American fiction—20th century—History and criticism. 2. Chaotic behavior in systems in literature. 3. Literature and science—United States—History—20th century. 4. Postmodernism (Literature)—United States. I. Title.
PS374.C4 C66 2002
813′.5409113—dc21

2001005650

British Library Cataloguing-in-Publication Data available

In memory of Louise Conte, and for my parents

Contents

Acknowledgments

All books are gotten out of other books, and this one is no exception. Even dutiful acknowledgments can't begin to tally the indebtedness to the thought and expression of others. My earlier work on seriality and proceduralism in postmodern poetry suggested that questions of contingency and structure, the interrelationship of order and disorder, were constitutive of contemporary aesthetics. Reading the fiction of John Hawkes (from whose novel *Travesty* I take my title phrase, "design and debris"), Kathy Acker, Robert Coover, Gilbert Sorrentino, Harry Mathews, John Barth, Don DeLillo, Robert Coover, and Thomas Pynchon, confirmed my impression that I was contributing to a single book on the poetics of chaos. However much I'd like to think it was my own curious pleasure that drew me to these novels, I discovered that each already contained their statement of a chaotics. In some instances the postmodern novelist displays a ready familiarity with contemporary science writing by James Gleick, N. Katherine Hayles, Stuart Kauffman, Benoit Mandelbrot, Ilya Prigogine and Isabelle Stengers, Michel Serres, or Norbert Wiener. One can find explicit references to cybernetics, hypertextuality, fractals, strange attractors, and complexity theory in novels by Barth, DeLillo, Pynchon, Italo Calvino, and William Gibson. But in the spirit of complementarity, one also finds that fiction writers such as Acker, Hawkes, Coover, and Sorrentino anticipate developments in the sciences of orderly disorder, speculating on the irregular nature of things before the dissemination of scientific studies. There was, in intellectual terms, no "two cultures" divide. "It all goes along together," as Pynchon says. "Parallel, not series. Metaphor. Signs and symptoms. Mapping on to different coordinate systems."

I am grateful to several colleagues at the University at Buffalo whose response to drafts of various chapters provided the encouragement necessary to assemble a book during the past six years. Robert Daly shared his wit and flawless recall regarding the work of our fellow Cornell University alumnus Thomas R. Pynchon. Raymond Federman invited me to teach Hawkes, Calvino, and Beckett in his seminar on experimental fiction, and at least some of that material survives in the pages of this book. Kenneth Dauber, Henry Sussman, James Bunn, and Mark Shechner provided exacting reviews of the entire manuscript. One cannot hope for any greater reward than to have such astute and demanding readers. Support from the Dean of Arts and Sciences, Kerry Grant, came in the form of a sabbatical in 1994 during which three chapters were first drafted. I also gratefully acknowledge a grant from the College of Arts and Sciences Dean's Publication Subvention Fund to defray the cost of copyright permissions. The Department of English sponsored a presentation of "The Superabundance of Cyberspace: Postmodern Fiction in the Information Age" as part of the Victor H. Johnson Lecture Series in 1998. My graduate students in several seminars devoted to Postmodern Fiction, Literary Theory, and Poetics were an invaluable resource over the past six years. The intense conversations across the seminar table, the revelations and intransigencies of critical reading and writing—as an act that one performs, as an art that one can master—have become too rare a commodity in our cable-ready, modem-connected information age.

Other readers offered critical advice on several chapters, not out of professional obligation but in unselfish support of an intellectual community. Martin Rosenberg and John Krafft made gifts of their encyclopedic knowledge of Pynchon's fiction and the related theoretical and critical literature. Doug Rice organized a memorable conference on Postmodern Piracy that featured Acker's work. Michael Joyce and Stuart Moulthrop provided generous responses to my essay on "The Superabundance of Cyberspace" at a panel on Multimedia Literature, organized by Midori McKeon for the Modern Language Association convention in 1997. Tom LeClair offered guidance on the "maximal" fictions of DeLillo, Pynchon, and Gaddis. My thanks to the other founding members of the Don DeLillo Society, especially John Duvall, Phil Nel, and Mark Osteen, whose conversation at conference panels and business meetings enriched my understanding of DeLillo's career.

Earlier versions of most chapters were presented as lectures or conference papers, during which attentive audiences completed the feedback loop of hypothesis and argument. I am especially grateful to Pil Dahlerup who arranged my Ph.D. seminar on postmodern literature and science at the Institute for Nordic Philology, University of Copenhagen, in 1996. Lars Saetre likewise deserves my gratitude for hosting me during subsequent lectures on "design and debris" at the University of Bergen, Norway. My thanks also to Laszlo Géfin for inviting me to speak on "Postmodern Fiction in the Information Age" in the Public Lecture Series at Concordia University, and to those who came to the auditorium in sub-zero weather in Montréal, Canada, in late

January 1997. Over the past few years, I have had the opportunity to present working drafts of chapters at the meetings of the Society for Literature and Science, Twentieth-Century Literature at the University of Louisville, and the American Literature Association. These gatherings became the necessary proving ground for my work. Versions of two chapters have appeared in the following journals: "'Design and Debris': John Hawkes's *Travesty*, Chaos Theory, and the Swerve," *Critique* 37 (1996): 120–38; and "Discipline and Anarchy: Disrupted Codes in Kathy Acker's *Empire of the Senseless*," *Revista Canaria de Estudios Ingleses* 39 (1999): 13–31. My thanks to the editors of these journals for permission to reprint.

I want to thank Katherine Hayles and Hank Lazer for their detailed evaluations of the final manuscript. Brian McHale's books on postmodern fiction were models for much of what I hoped to accomplish in *Design and Debris*. As a reader for the University of Alabama Press he provided invaluable direction for the completion of the manuscript. Charles B. Harris served as a reader for the chapter on Hawkes published in *Critique*, and later as a reader of the manuscript for the University of Alabama Press. His evaluations were decisive in focusing the themes and contents of the book in its present form. I am grateful to Curtis Clark, Assistant Director and Editor-in-Chief of The University of Alabama Press, for his patience and encouragement while the manuscript reached completion. My appreciation also extends to the staff of The University of Alabama Press for their assistance and vigilance in several stages of manuscript preparation, especially Dan Waterman and Mindy Wilson.

Gilbert Sorrentino has been my cicerone through so much postmodern writing that it is difficult for me to imagine my work in this field without his direction. Diane Middlebrook has proved time and again that an abiding friend and mentor can be united in one person. Robert Basil has been my "brother in the work" over the years. Joy Leighton understood that what we demand of ourselves is also demanded of others; she has been my true partner in all of this. Finally, my deepest gratitude to my family, especially my parents Ralph and Ann, without whom there could be nothing.

Design and Debris

I

Being in Uncertainties

Orderly Disorder in Postmodern American Fiction

> Yet so blind are we to the true nature of reality at any given moment that this chaos—bathed, it is true, in the iridescent hues of the rainbow and clothed in an endless confusion of fair and variegated forms which did their best to stifle any burgeoning notions of the formlessness of the whole, the muddle really as ugly as sin, which at every moment shone through the colored masses, bringing a telltale finger squarely down on the addition line, beneath which these self-important and self-convoluted shapes added disconcertingly up to zero—this chaos began to seem like the normal way of being, so that some time later even very sensitive and perceptive souls had been taken in: it was for them life's rolling river, with its calm eddies and shallows as well as its more swiftly moving parts and ahead of these the rapids, with an awful roar somewhere in the distance; and yet, or so it seemed to these more sensible than average folk, a certain amount of hardship has to be accepted if we want the river-journey to continue; life cannot be a series of totally pleasant events, and we must accept the bad if we also wish the good; indeed a certain amount of evil is necessary to set it in proper relief: how could we know the good without some experience of its opposite?
>
> —John Ashbery, "The System"

Each discipline in its own terms, each with its own methodology, has sought to explain the emergence of order in a world in which entropy dictates an inexorable tendency towards disorder. The astronomer charting the contraction of a protostar from interstellar gas, the theologian in pursuit of the first mover, the biologist parsing the evolution of a species, and the novelist in search of a unitary and previously unspoken idea: each asks, Why are there orderly structures and complex living beings when inchoate shapes and disorderly activity should prevail? No work of fiction could provide a satisfactory answer to this question, but the novelist in every creative act searches for the principles by which order might arise out of the disorder of the materials. Such coherence can't be coaxed or forced, but is allowed to be made manifest in the process of the writing. A novel's coming into being enacts on its limited scale the same implausible birth of a complex and sustaining world. An author experiences this emergence of order—though it resists explanation—and participates in its formation, its economy of means, and its grace.

And yet in disorder lurks the figure of reality. To isolate the regular from the irregular, the shapely from the disfigured, or the coalescent from the decaying—as art, philosophy, or science has sometimes done—provides no adequate portrait of the world. The novelist must look into the turbulent crowd to find the unanticipated face of composure. Disorder is the more prevalent condition, the more probable state of affairs; one ignores it at the risk of an utter and fruitless detachment. So too the possibility of a new order arises not from stasis but out of tumult. Here artists have an advantage over scientists. Though both professions are driven by a desire for discovery, the artist is not charged with formulating laws or generalities. Rather than ascertain the properties that are common or universal to phenomena, the artist searches for those inimitable particles that distinguish an object. In this the artist may first recognize the coruscation of the valuable against the lack of differentiation and homogeneity of a world governed by entropy. The artist also refuses to exclude from consideration those phenomena that do not submit readily to some process of normalization. The novels that I discuss in this book express the dual purpose of realizing the emergence of pattern while resisting the regularization of experience into the soluble and coherent alone.

Postmodern fiction enacts the interrelation of order and disorder in its world-making and in the inventive form of the novel. This study examines the coeval development of nonlinear narrative and the concept of orderly disorder as it has been promulgated by scientific theories of chaos and complexity since the 1970s. The publication of some works of fiction that are both steeped in disorder and illustrative of self-organization, such as Robert Coover's *The Universal Baseball Association, Inc.* (1968) and John Hawkes's *Travesty* (1976), precedes the appearance of major studies in chaotics such as Benoit Mandelbrot's *The Fractal Geometry of Nature* (1983) and Ilya Prigogine and Isabelle Stengers's *Order Out of Chaos: Man's New Dialogue with Nature* (1984). And yet it would be inappropriate to say that these novels "anticipate" discoveries in the fields of fractal geometry or nonequilibrium thermodynamics. Conversely, John Barth states that the feedback loops in *The Last Voyage of Somebody the Sailor* (1991) are an expression of "chaotic-arabesque Postmodernism" (*Further Fridays* 289); Thomas Pynchon punctuates his treatment of televisual media and conservative politics in the 1980s in *Vineland* (1990) with an awareness of fractal forms in nature; and William Gibson and Bruce Sterling's retrograde fantasy of Victorian computing, *The Difference Engine* (1991), is replete with references to complex systems and the "period-doubling route to Chaos" (211). These explicit acknowledgments of chaotics in postmodern fiction, however, should not be considered attempts to "adapt" or metaphorize scientifically verifiable theories in literary works. There is little to gain in debating either an originary or derivative source for chaotics in postmodern culture, nor do I see the need to extend Jean-François Lyotard's indictment of the cultural imperialism of denotation in scientific discourse and its subordination of the "fuzzy," connotative functions of narration (25–27). Rather this study contends that there has been a homologous development in which the two disciplines move independently but to shared convictions regarding the nature of chaos. Correspondences between the self-similarity of tropes in a literary form and in a fractal are not accidental, but the refraction of a common observation in each discipline. The similarities between postmodern fiction and nonlinear dynamics, however, should not obscure the distinctive demands that are integral to each discipline.

Whatever specific correlation between the forms of orderly disorder and postmodern fiction may be argued in the following chapters, let me suggest that there are two essential conditions found in both disciplines: uncertainty and unpredictability. In his recent treatment of irreversibility and the "arrow of time," *The End of Certainty*, Ilya Prigogine, who won the Nobel Prize in 1977 for his research in nonequilibrium thermodynamics, argues that there has been a shift from the classical science that emphasized order and stability to—he doesn't quite call it "postmodern"—the physics of nonequilibrium processes and unstable systems. Whereas the laws of nature in the classical view were expressed as certitudes, the new physics, incorporating irreversibility and instability, expresses the laws of nature in terms of possibilities and probabilities (4). We have reached, he declares, the end of certainty. Chaotic systems are subject to "sensitive dependence on initial conditions." Because the origin of the system cannot be determined with infinite precision, it becomes impossible to predict the subsequent state of the system. In nonlinear dynamics even the closest of trajectories will diverge exponentially over time. Lyotard concurs, responsible as he is for proposing a "postmodern science" that would be concerned with the "search for instabilities" (53–60). He challenges the representation of scientific knowledge as the totality of knowledge and its suppression of narrative. The turn toward unpredictability in postmodern science establishes an affinity with narrative that classical science, in its demand for deterministic results, could not permit. As the gaze of science awakens to conditions of unpredictability and uncertainty, the possibility of an alliance with postmodern theory and fiction arises.[1] Postmodern fiction dwells at the interstices between order and disorder, certainty and uncertainty, verifiability and supposition.

In all of the novels discussed in this study I find the interrelation of design and debris. But for my own pleasure in establishing some order, the chapters have been arranged to present an alternation of emphasis between novels of design and novels of debris. Hawkes's *Travesty*, Gilbert Sorrentino's *Pack of Lies* (1997), and Coover's *Universal Baseball Association* are works in which an undisclosed pattern becomes apparent in disorder; they are novels that reveal an immanent design in the fractious conditions they describe. They are homologous in form to the deeply encoded structures that are the subject of nonlinear dynamics. Kathy Acker's *Empire of the Senseless* (1988), Don DeLillo's *White Noise* (1985), and Pynchon's *Gravity's Rainbow* (1973) are works for which the disorderly confusion of the phenomenal world, the disruption of staid and resolved structures, becomes a requirement for the emergence of a new and more desirable order. They revel in the rich debris that is the source of self-organization and complexity. Further pairings in this mode are limited only by the space available to me, but the few couplings presented here should be sufficient to illustrate my position. Linda Hutcheon remarks in *The Politics of Postmodernism* that there is an inherent duplicity in postmodernism as a result of its ironic cast (1). Thus one finds in discussions of postmodernism the prevalence of the conjunction "both/and" rather than the disjunction "either/or." If postmodernism can be described by the ironic and unresolved pairings of representation and self-reflexivity, complicity and critique, continuity and rupture, so it must be with my treatment of design and debris in the novels. How these characteristics are read together will finally express a chaotics of postmodern fiction.

The Paradigm Shift in Postmodern Fiction

The transition from modernism to postmodernism has been described in historiographic terms as an incremental and continuous change from one period with its defining events and cluster of authors and theorists to another. But for the present discussion I would like to examine the transition to postmodernism in the terms of what Thomas S. Kuhn has famously called a "paradigm shift." Although historiographic and dialectic readings of postmodernism have proven invaluable,[2] there are several characteristics of Kuhn's treatise on paradigms that are distinctly apparent in the shift from the modern to the postmodern period. He defines a scientific revolution as an episode in which "an older paradigm is replaced in whole or in part by an incompatible new one" (92). Upsetting the dignity of scientific institutions, Kuhn contends that the majority of scientists remain for some time after the introduction of a new paradigm committed to the incremental processes of "normal science" (in which they continue to address legitimated problems and methods), withholding acknowledgment or acceptance of model-shattering discoveries (24). Bringing about a paradigm shift, in which new and incommensurable beliefs displace an older methodology, is a task often "restricted to a narrow subdivision of the scientific community" (92). By distinguishing between the "mopping up operations" (24) that occupy most scientific careers and the extraordinary discoveries that precipitate scientific revolution, Kuhn attributes an extra-liminal perception to those who pursue theories incompatible with the familiar things as they are. These individuals—Galileo, Copernicus, Curie, Einstein—become recognized as geniuses. Like Wallace Stevens's man with a blue guitar, in part a homage to Pablo Picasso's imaginative and transforming vision, these great figures of discovery play "A tune beyond us, yet ourselves, . . . Of things exactly as they are" (*Collected Poems* 165). Modernism, more so than other literary-historical periods, extols the figure of the inventive genius: artists such as Picasso, Gertrude Stein, and James Joyce possess the ability to apprehend the world in other than its conventions, beyond the threshold of the familiar, in a mode that is more exact, even as it offends bourgeois respectability and the norms of artistic practice. Kuhn argues that the incompatible or anomalous aspects of the contending paradigm trigger a crisis in "normal science." One finds among modern writers and artists an explicit commitment to the pursuit of the incommensurable, mold-breaking practice.[3] William Carlos Williams, in a review of George Antheil's cacophonous *Ballet méchanique* (for pianos, percussion, and airplane propeller) in 1927, remarks, "Everything new must be wrong at first since there is always a moment when the living new supplants that which has been and still is right and is thus sure to be wrong in transit, or until it is seen that that which was right is dead" ("George Antheil" 60). Williams recognizes that the modernist pursuit of the new inevitably results in anomalies when seen in the context of an existing paradigm until that moment when the shift to a new paradigm occurs.

Vindication for artist or scientist occurs when the incompatibilities between competing paradigms force a break or rupture—not unlike that attributed to postmodernism in its displacement from modernism—in which the destabilized beliefs of the old paradigm give way to the new and now strangely acceptable paradigm. At these times, Kuhn argues, "the scientist's perception of his environment must be re-educated—in some familiar situations he must learn to see a new gestalt. After he has done so the world of his research will seem, here and there, incommensurable with the one he had inhabited before" (112). Paradigm shifts, when they occur, force a pervasive revision of the scientist's world. Like tectonic plates grinding along a fault line in opposite directions that shift suddenly in an earthquake, the slippage of paradigms comes without warning, a gestalt-quake. Indeed, Kuhn regards "scientific development as a succession of tradition-bound periods punctuated by non-cumulative breaks" (208). In this thesis he has the company of Stein who says, "the creator of the new composition in the arts is an outlaw until he is a classic, there is hardly a moment in between" (514). Although the abruptness of paradigm shifts can be identified in most successions of periods—including Virginia Woolf's announcement that "in or about December, 1910, human character changed" (91) or Charles Jencks's claim for the "sudden demise" of Modern architecture (*Post-Modernism* 27) with the demolition of the Pruitt-Igoe housing project in St. Louis on the afternoon of July 15, 1972[4]—the "non-cumulative break" plays a special role in postmodernism. Modernism cultivated as doctrines the individual genius in alienation from society, the autonomy of the subject, and the auto-referentiality of the work of art. The crisis of language that arises in modernism can be thought of as one example of the impossibility of a mutually agreed upon meaning for terminology between contending paradigms. Postmodernism adopts a "model of rupture" (Hutcheon 27) that is not predicated on individual alienation within a capitalist, classist, war-machine state but rather on an irremediable breach in the cultural construction. Lyotard has famously defined the postmodern as "incredulity toward metanarratives" (xxiv). In his formulation postmodernism signals the "delegitimation" (37) of the supposedly universal truths of western humanism such as political emancipation, technological progress, philosophical consensus, imperialism, logocentrism, and the patriarchy. Postmodernism's break with the putatively transcendent beliefs of the Enlightenment represents the sort of shift whose repercussions affect all aspects of the culture and are not limited to the discourse or practice of a single discipline. The penchant for "dedoxification" (Hutcheon 3), "decreation/deconstruction" (Hassan 91), "denaturing" (Hayles, *Chaos Bound* 265–95), and "delegitimation" in postmodernism confirms that a break with widely held beliefs is not simply an unavoidable effect of shifting paradigms but a dominant motive in postmodernism.

Theorists and critics have advanced several candidates for the constitutive shift from the poetics of modernism to that of postmodernism. Brian McHale suggests that there has been a "shift of dominant from problems of *knowing* to problems of *modes of being*—from an epistemological dominant to an *ontologi-*

cal one" (*Postmodernist Fiction* 10); that is, from questions of interpretation to strategies of world-making. Hutcheon singles out the shift from "the exclusion of mass culture" in modernism to "postmodernism's renegotiation of the different possible relations (of complicity and critique) between high and popular forms of culture" (28). She in turn critiques Fredric Jameson's distaste for postmodern art's appropriation of commodity culture and his position that "market capitalism begat realism; monopoly capitalism begat modernism; and therefore multinational [or consumer] capitalism begat postmodernism" (25; see Jameson 35–36). Jameson acknowledges from the start of *Postmodernism, or, The Cultural Logic of Late Capitalism* that the case for postmodernism's "existence depends on the hypothesis of some radical break or *coupure*, generally traced back to the end of the 1950s or the early 1960s" (1). Each thesis recognizes a "non-cumulative break" with the modernist paradigm, although the anomalous emphases and timelines suggested by these examples demonstrate that we are still in a phase of paradigm-building with regard to the poetics of postmodernism.

I contend that in the arts and sciences a paradigm shift occurs in postmodernism in the conception of the relation between order and disorder. In modernism the function of the artist is to impose order and coherence on a disorderly, random, and inchoate world. The "extraordinary science" of this period—Albert Einstein's Special Theory of Relativity (1905) that discards the concept of absolute motion, Werner Heisenberg's Uncertainty Principle (1927) that implicates the observer as a presence in any system of measurement, and Niels Bohr's Principle of Complementarity (1927) that treats the logical contradictions of light as wave and photon as complementary—irrefutably challenges the deterministic logic of the natural world that had held sway since Isaac Newton published his *Principia* (including his three laws of motion) in 1687. The onset of relativism and indeterminacy undermines the linear reasoning of proportional causality, continuity, and objectivity that supported realism.[5] Modernism expresses the shock of an apparent collapse of intrinsically ordered systems of the physical world and the assumptions of social behavior and forms of communication that depended on the natural order of things for their legitimacy. This blow to determinism liberates the artist from representationalism, promotes a self-conscious and "more intuitive" aesthetic, and permits art to fulfill itself on its own terms (Bradbury and McFarlane 25). Recognizing that the "world, reality is discontinuous until art comes along," the modern artist responds to the crisis by making that discontinuity and its license to expression subject to "the aesthetic system of positioning" in the work of art (Bradbury and McFarlane 25). I consider the compulsion to order in modernism essential to its aesthetic system, and it finds its expression in some of the most memorable phrases of the period. In "The Idea of Order at Key West" Stevens invokes the "maker's rage to order" the turbulence and flux of the world (*Collected Poems* 130).[6] Near the close of his notoriously difficult rendition of the dissolution of modern life, *The Waste Land*, T. S. Eliot asks, "Shall I at least set my lands in order?" and disconsolately offers only "These fragments I have shored against my ruins" (50).[7] The aspiring artificer, Stephen Daedalus, in James Joyce's *Portrait of the Artist as a Young Man*, states that the modern "artist, like the God of the creation, remains . . . invisible, refined out of existence, indifferent, paring his fingernails" (215), a dictum of impersonality and restraint that is remarkable, as Lynch reminds Daedalus, in the midst of the Irish turmoil. Joyce alludes here to a letter of Flaubert in which he calls his novel *Madame Bovary* a work of complete invention without debt to reality.[8] The "Flaubertian dream of an order in art independent of or else transcending the humanistic, the material, the *real*" (Bradbury and McFarlane 25) is one source of the modern commitment to formalism in all the arts. So Eliot takes it upon himself to rebut Richard Aldington's charge that Joyce's *Ulysses* was "an invitation to chaos" by suggesting that Joyce's mythical method was "simply a way of controlling, of ordering, of giving a shape and a significance to the immense panorama of futility and anarchy which is contemporary history" ("*Ulysses*, Order and Myth" 177). The modern artist as described by Eliot's classicist, Joyce's creator, Stevens's maker, or Ezra Pound's "factive personality," is a fabricator of order. These writers are aware that the patterns they impose are tenuous assertions of their creativity against the darkness and dissolution that surrounds them. Thus the appeal to genius—the one who acts with autonomy, possesses an elitist sensibility, and figures as the *primum mobile* of creation and invention—validates the modern artists' shaping of an inchoate world in their own image.

As the dominant motive in postmodernism constitutes a "break" in the condition of knowledge arising from a renunciation of several foundational myths of western culture, it follows that the postmodern artist expresses an affinity for—rather than an aversion to—forms of disorder. Recognizing an epistemic shift in the relation of order and disorder, the postmodern artist no longer regards disorder as a solely destructive or irrational state, as if it were forever wedded to civil insurrections or natural catastrophes (such as the revolution of 1871 in Paris or the eruption of Vesuvius that sealed Pompeii in 79 AD, events which tend to inordinately discomfit the bourgeois). Postmodern theories such as the "deterritorialization" of Gilles Deleuze and Félix Guattari and the "delegitimation" of Lyotard actively seek to destabilize orderly institutions and revel in the unpredictability that results as a forum for unrestricted play and the possibilities of discovery. The postmodern writer has become, to borrow another phrase from Stevens, a "connoisseur of chaos" (*Collected Poems* 215). If human orders are merely personal and contingent, offering no proof of a general orderliness, and if the disorderliness of nature may result in a thing of beauty beyond art, then the philosophical propositions of an inherent opposition of order and disorder, or between culture and nature, meet with a barrage of exceptional evidence. There remains the very human compunction to realize an absolute distinction between order and disorder. There is also a very human fear that these two states are fundamentally indistinguishable or in constant exchange. Confronting this issue has equally important consequences for the author, the theorist, and the scientist in postmodernism. The crucial turn of mind occurs in the appreciation of order as the possible emanation of disorder, and chaos as one possible result of an overly stringent order—the process by which the one becomes the other. Not the identification of the two; not the assertion of a null set that divides the two into camps of exclusion. The composer John Cage acknowledges the essential role of postmodern artists in life and art: "Our intention is to affirm this life, not to bring order out of a chaos nor to suggest improvements in creation, but simply to wake up to the very life we're living" (*Silence* 95). Cage's acceptance of unpredictability and his unwillingness to make the imposition or discovery of order the determinant of art mark the shift from modernism. The fascination for the postmodern artist is not that order and disorder are unchanging like statuary, but that they are always contingent and complexly interrelated, like the shadows of chalk marks and dust on a blackboard. The postmodern writer proposes a text and a world engaged in the continual process of invention and disruption, the making of meaning and the free play of signifiers, and an ever-shifting interpenetration of figures of order and chaos.

In this activity the postmodern writer has been especially willing to embrace what Lyotard considers the new subject of postmodern science, the search for instabilities. In *The Postmodern Condition*, Lyotard asserts that science has reached a "crisis of determinism" (53). The conduct of normal science relies upon an idea of performance with a linear and predictable relation of input to output. The thermodynamic model of performance implies a highly stable system such that, if all the variables of the initial state of the system were known, its state at any future point could be accurately predicted. But quantum mechanics and nonlinear dynamics cast the discernment of the initial state of the system into uncertainty. The radioactive disintegration of radium, while calculable in the aggregate by statistical probability, is unpredictable for any single nucleus. Besides this shift from determinism to unpredictability, stable to unstable systems, Lyotard notes the contributions of mathematician Benoit Mandelbrot, who shows the limitations of a science based largely on forms of regularity. In the language of geometry, Mandelbrot explains, irregular "curves that have no tangent are the rule, and regular curves, such as the circle, are interesting, but quite special" (cited in Lyotard 58). Science has made the special case of soluble, linear equations its normative concern, relegating as exceptional cases the irregular and nonlinear forms —"the contours of a floccule of soapy, salinated water" (58)—that are in actuality more prevalent.[9] Mandelbrot's "fractal" geometry of self-similar forms such as the Koch curve that emulate the irregularity of a coastline or snowflake, and René Thom's catastrophe theory that mathematically analyzes discontinuities and unstable systems, represent an epistemic shift, a change in the condition of knowledge. Lyotard concludes, "Postmodern science—by concerning itself with such things as undecidables, the limits of precise control, conflicts characterized by incomplete information, '*fracta*,' catastrophes, and pragmatic paradoxes—is theorizing its own evolution as discontinuous, catastrophic, nonrectifiable, and paradoxical. It is changing the meaning of the word *knowledge*, while expressing how such a change can take place. It is producing not the known, but the unknown" (60). The paradigm shift of postmodern science represents what the poet Robert Duncan called "the opening of the field." Science no longer limits its attention to objects and conditions that have undergone regularization by the human mind. Instead, it turns its attention to the vast array of natural phenomena—"the coast of Brittany, the crater-filled surface of the moon, the distribution of stellar matter, the frequency of bursts of interference during a telephone call, turbulence in general, the shape of clouds" (Lyotard 58)—whose instability, irregularity, and beauty surround us; a shift from nonlinearity as a grotesque exception in a deterministic universe, to a world in which orderly disorder is plentiful and predictability a rarity.[10] The postmodern writer both anticipates and fervently illustrates theories of chaotic behavior and unpredictability before their adoption by the academy as a legitimized narrative.

Interdisciplinarity and the Culture of Chaotics

Ihab Hassan's table of distinctions between modernism and postmodernism assigns a preoccupation with "centering" (91) to the literature of the early twentieth century, though that centering is often expressed in terms of crisis, as the heroic effort of the artist to assert coherence against the disorderly forces of disruption. In "The Second Coming," W. B. Yeats anxiously prophesies a "widening gyre" in which "Things fall apart; the centre cannot hold; / Mere anarchy is loosed upon the world" (187). Pound agonizes of his epic poem, *The Cantos,* and of the panorama of modern culture, "I cannot make it cohere" (Canto 116, 810). Eliot retreats to a *hortus inclusus* in "Burnt Norton." Each of these modern poets, as Michael North argues in *The Political Aesthetic of Yeats, Eliot, and Pound,* adopts a political aesthetic whose concern for a centering control approaches fascism. Postmodernism distinguishes itself by an affinity for "dispersal." Like droplets of oil on water, the work of postmodernism diffuses itself to the widest extent possible. The limitless connection of the rhizome (Deleuze and Guattari 7), schizophrenia as a breakdown of the signifying chain (Jameson 26), "grazing" in restaurants, browsing on the Internet, multiculturalism, world Anglophone literatures, the global economy, and the telecommunications nexus, are figures of postmodern dispersal. Instead of the condensations of lyric poetry, postmodernism deploys the scattering of textual scraps in the cut-up novels of William Burroughs, the saturation of electronic media into every crevice of the American environment in DeLillo's *White Noise,* or the dissipation of vehicular collisions in J. G. Ballard's *Crash.*[11] One reads these works not as the failure of an authoritative individual to assert unity and preserve order, but—with apologies to the Children's Television Workshop—as an indiscriminate "sharing," a pluralism, or an always-open international *bourse* of cultural exchange.

The dynamic of dispersion in postmodern culture percolates through all fields that inquire into the nature of disorder. The professional disciplines of the arts and sciences, and the realm of popular culture, are equally affected by this paradigm shift. Gilles Deleuze and Félix Guattari propose a "nomad" science that contests the centralization of a State science whose program supports the "war machine" of the military-industrial order. But the unfettered dynamic of nomad science—in every historical period—persists despite the formalization of its inventions and the marginalization of its method by the State in the figure of the institute or the engineer. Since the 1970s theories of chaos and complexity have developed as a form of nomad science. Whereas State science needs the discipline of hydraulics to subordinate the force of water to conduits and pipes for its own use, nomad science studies the "turbulence across a smooth space [the flood], in producing a movement that holds space and simultaneously affects all of its points, instead of being held by space in a local movement from one specified point to another" (363). The shift from the striated model of State science to the smooth model of nomad science occurs repeatedly in postmodernism.

Deleuze and Guattari find the figures of stabilizing rhythm and desultory disorder interwoven in even the mundane forms of popular culture. A song with its refrain represents a stable "center in the heart of chaos," to alleviate the anxiety and confusion of life. But the song has already begun to skip in its groove: "it jumps from chaos to the beginnings of order in chaos and is in danger of breaking apart at any moment" (311). It is from chaos, they claim, that milieus and rhythms are born. But milieus are in turn "open to chaos, which threatens them with exhaustion or intrusion. Rhythm is the milieus' answer to chaos. What chaos and rhythm have in common is the in-between—between two milieus, rhythm-chaos or the chaosmos[12]. . . . In this in-between, chaos becomes rhythm, not inexorably, but it has a chance to. Chaos is not the opposite of rhythm, but the milieu of milieus" (313). Deleuze and Guattari challenge the exclusive opposition between pattern and disorder. They seek a "nonlocalizable, nondimensional chaos, the force of chaos, a tangled bundle of aberrant lines" (312). The postmodern milieu of dispersion invites rather than represses such a rhythm-chaos.

N. Katherine Hayles distinguishes between the science of chaos, or nonlinear dynamics, and the broader cultural phenomenon of "chaotics," a term which she attributes to Hassan. Chaotics signifies "certain attitudes toward chaos that are manifest at diverse sites within the culture" (*Chaos and Order* 7).[13] Rather than the promulgation of chaotics from a single authoritative source, each cultural site emanates new conceptions of disorder and complexity that are not reliant upon, but may be cognate with, the others. As Leni Pökler in Pynchon's *Gravity's Rainbow* explains to her "cause-and-effect man," Franz Pökler: "It all goes along together. Parallel, not series. Metaphor. Signs and symptoms. Mapping on to different coordinate systems" (159). There is now a diaspora of theories of chaos and complexity in disciplines as diverse as economics, meteorology, music, politics, biology, and literature. The economist Brian Arthur's study of increasing returns and the "lock-in" of standards such as the QWERTY keyboard or the VHS videotape due to positive feedback shows "how chance events work to select one equilibrium point from many possible in random processes" (Waldrop 46). The meteorologist Edward Lorenz's attention to turbulence in weather systems led to his proposal of the Butterfly Effect—or the more prosaic "sensitive dependence on initial conditions"—in which small causes in nonlinear systems can cascade upward into large effects due to positive feedback. The composer John Cage describes his *Fontana Mix* (1958) for magnetic tape as "a composition indeterminate of its performance." The freedom to superimpose transparent sheets of directions regarding sound sources, their mechanical alteration, changes in pitch and duration, and the looping or splicing of tape allows the musician to render many differing performances.[14] The model of dispersion, the acceptance of unpredictability, and the interrogation of disorder in postmodernism are likewise not locked in a linear, causal logic but are shared aspects of an complex, emergent system.

This "crosstalk" (the coupling across transmission circuits) between the disciplines of the sciences and the humanities makes chaos theory exemplary of a postmodern turn toward interdisciplinary knowledge and polymathy. Ancillary to the shift in attitude toward disorder, the interdisciplinarity of postmodernism represents a paradigm shift from the intensive specialization that governed discourse in modernism.

In the early twentieth century the fields of knowledge—the physical sciences, the humanities, the social sciences—underwent a rapid specialization. It became increasingly difficult for a well-educated individual to attain a professional level of knowledge outside of a single concentration of studies. The historian and statesman Henry Adams illustrates the strain of specialization and the resulting rift between the disciplines in his account in *The Education* (1907) of his informal exchanges with the physicist Samuel Pierpont Langley, who was in residence at the Smithsonian Institution in 1900. Langley "had the physicist's heinous fault of professing to know nothing between flashes of intense perception. . . . Rigidly denying himself the amusement of philosophy, which consists chiefly in suggesting unintelligible answers to insoluble problems, he still knew the problems, and liked to wander past them in a courteous temper, even bowing to them distantly as though recognizing their existence, while doubting their respectability" (377). Langley's courteous disdain for philosophy and, for his part, Adams's confession that the concepts of modern science were already beyond his understanding or inquiry, represent the quiet tearing of an intellectual coat. Adams had pursued history for its revelations of "direction or progress," and now he found only the random motion of kinetic atoms; he had sought unity and simplicity, but found multiplicity and complexity. Finally, faced with the electric dynamo at the Paris Exhibition of 1900, he was forced to relinquish the historian's attachment to cause and effect. The scientist "had entered a supersensual world, in which he could measure nothing except by chance collisions of movements imperceptible to his senses, perhaps even imperceptible to his instruments, but perceptible to each other. . . . Langley seemed prepared for anything, even for an indeterminable number of universes interfused—physics stark mad in metaphysics" (381–82). With his "historical neck broken by the sudden irruption of forces totally new," Adams worships the Dynamo in the absence of understanding, just as the Incan priests conferred godly status to the first conquistadors on horseback who appeared in the village.

The instigation of what C. P. Snow would call the "two cultures," the literary and the scientific, in his 1959 Rede Lecture provides a platform for the specialization of disciplines that further divides the fields of knowledge. At almost the same moment that Adams's neck is broken, there are significant advances in psychology and philosophy that appropriate part of the interior domain that might have been left to the arts after the sack of the supersensual by the sciences. William James's *Principles of Psychology* (1890), Sigmund Freud's *The Interpretation of Dreams* (1900), Henri Bergson's *Time and Free Will* (1889), aspire to the very respectability and seriousness of the labcoat that Langley had denied to disciplines that deal with the mind's self-analysis. The arborescence of disciplines, the formalization of modes of inquiry and the jargon that attends them, the inevitable hierarchizing that granted privilege to the sciences (and those disciplines that aspired to scientific status) constitutes a pronounced fragmentation at the last turn of the century—the division of humanist inquiry into fields of specialization.

This flourishing of disciplinary specialization coincides with the period of modernism in the arts.[15] Each field generates a terminology whose assigned meanings are specific and exclusive to its discipline-of-origin. In Henry Reed's "Naming of Parts" from *Lessons of the War* (1946), an infantryman recalls his drill sergeant's instructions on the handling of firearms that begins, "Today we have naming of parts" (49). Like the sergeant's insistence on the proper name and use for the "lower sling swivel" or the "safety-catch," the specialization of the intellectual project in modernism results in a loss of shared terminology between disciplines.[16] Literary modernism follows the trend in science toward reductionism, the analysis of systems in terms of their constituent parts. Physicists smash atoms into ever smaller elementary particles, to which they give the names quarks, leptons, and mesons.[17] Neurologists examine the most complex organ, the brain, in terms of its circuitry of neurons. Literary critics of the mid-century outline the pathetic fallacy, symbolic logic, and the function of metonymy. In pursuit of specialization, modernist writers emulated various aspects of scientific disciplines in their own methodology: James informs the psychological typologies of early Gertrude Stein; Freud's treatment of desire and James's "stream of consciousness" affect the narratology of Joyce, William Faulkner, and John Dos Passos; Bergson's philosophy of process influences Stevens's quest for poetic truth. As a physician and poet Williams complains that his undergraduate medical training at the University of Pennsylvania left him desperately unprepared (in contrast to his classmate, the young aesthete Pound) for a career as a writer. The penchant in modern music, art and literature for "difficulty" is itself a sign of the demand to specialization. Eliot—whom Snow brands as the "archetypal figure" of the literary intellectual for his enervated view that the world was suffering an entropic decline (5)—argued that difficult poetry was necessitated by the "variety and complexity" of modern civilization. But in their "difficulty" literature and the arts were expressing their desire to be treated as specialized modes of knowledge comparable to the physical sciences. An analogous movement occurs in structural linguistics and the New Criticism during the 1930s and 1940s as literary criticism was based in rules generated by language functions and structures rather than humanistic discourses of theme, biography, or creative intentions. Modernist literature sought to bring its argument to the ground of scientific discipline rather than marginalize itself as an entertaining conundrum or pleasant pastime.

Let us imagine for the moment that Snow's "two cultures" were those of the male and female genders, an emphasis in modernism that Hassan identifies as the "Genital/Phallic" (91). In this dichotomy it is not difficult to associate a gender with the "heroic" age of modern physics.[18] The orientation in postmodernism, however, has shifted to the "Polymorphous/Androgynous." Rather than propose a consummated union of literature and science that would leave them always a dissatisfied "other," postmodernism has achieved a "perverse" multidirectionality in its attentions—a polymathy—that is constantly shifting in form and indiscriminately appropriative in manner. The polymathic postmodern does not make a dilettantish inquiry into a few favored terms, but aspires to an interdisciplinarity that presumes a shared discourse, the diffusion of theoretical concepts into all quadrants of society, and the sense of an integrated feedback loop among the disciplines. Whereas modernism exerted itself in the naming of parts, postmodernism aspires to a curriculum of the whole. Thomas Pynchon, admitted to Cornell University to study engineering physics during the Sputnik era of the late 1950s in which Snow writes, responds to Snow's charge that literary intellectuals were "natural Luddites" (23).[19] Snow's social critique focused on the failure of communication between the literary and scientific communities. Twenty-five years later, Pynchon observes that "we have come to live among flows of data more vast than anything the world has seen" ("Luddite" 1). The dissemination of information, particularly through the use of the computer, has profoundly and in an egalitarian manner changed the culture of knowledge from the elite enclaves gathered at the Cambridge High Tables described in Snow's essay. Pynchon counters that "Demystification is the order of our day," and one regards with suspicion people who "hide behind the jargon of a specialty or pretend to some data base forever 'beyond' the reach of a layman" ("Luddite" 1). Now, forty years after Snow's debut and fifteen years after Pynchon's rejoinder, the Internet has become an enormous, and inexpensive, swap meet for information. The development of an information society coincides with the political elevation of multiculturalism with its equanimity toward ethics and values regardless of origin. One can no longer speak of a "two-cultures quarrel" when there are now so many cultures for which the exchange of values and ideas is inherently acceptable.

The shift from specialization in discrete disciplines to the polymathy of postmodernism finds an analogy in the field theories of physics. In place of a deterministic model in which the clockwork structure of the universe could be reduced to a set of component parts and then reassembled without damage to its functionality, field theory proposes a dynamic model of the universe whose whole is always something more than the sum of its parts. The field theory of physics, by which bodies separated in space exert gravitational, electric, or magnetic force on each other, confirms that there can be no solitary action, no autonomy or detachment of parts, and hence no artistic expression that occurs without relation to other actions, states, and processes. Applying such concepts to his poetics and metaphysics, Robert Duncan proposes in his volume of poems, *The Opening of the Field* (1960), and in essays such as "Towards an Open Universe," the interrelation or "rime" of fields of knowledge. Physicist Erwin Schrödinger's definition of living matter as that which evades the decay to equilibrium in *What is Life?* Anton Webern's twelve-tone method of musical composition, and the gnostic belief in hidden connections, Stuart Kauffman's "order for free" in the evolution of complex systems, are cognate processes of creation that are not modeled solely on the biophysical order, but serve as confirmation of the natural order that encompasses all fields. The appeal of the "open universe" lies in the continual exchange of material that allows no thought, no process, and no form to remain truly isolated. Duncan follows Charles Olson's theory of "composition by field" in his manifesto for an open-form poetics, "Projective Verse" (1950). A field theory of composition sets aside the modernist hierarchy of authorship for a postmodern interrelation (or parataxis) of the author, the text, and the reader.[20] These literary works record an epistemic shift from a closed universe of increasing disorder and homogeneity to an open universe that sustains complexity and heterogeneity. Postmodern resistance to the autoteleology of modernist texts, in which literature closes itself off from historical, biographical, and cultural interpretation, leaves the way open to the ideas and discourse of other-than-literary disciplines. No longer constrained by an ideal of textual autonomy espoused by modernist criticism, postmodern literature engages in a heightened degree of textual interactivity—between the "two cultures" of literature and science, and between author and reader—culminating in the infinite connectivity of hypertext fiction.

Orderly Disorder in Postmodern Fiction

A connoisseur is an individual of discriminating taste; one who recognizes the subtlest variations of character and quality; one who appreciates finesse. When Wallace Stevens speculates that one could become a "connoisseur of chaos" (*Collected Poems* 215), he establishes a coterie of twentieth-century writers, philosophers, and scientists who see chaos as something more subtle than the binary opposite of order. This opposition between chaos and cosmos is rooted deeply in Western philosophy; one might say that disfavor expressed toward chaos is the original prejudice, and thus not easily dislodged. In the cosmogony of Hesiod and for the pre-Socratic philosophers, Χάος (Chaos) is the "gaping void" or formless abyss out of which the primordial matter of the universe evolves. It is the first state or infinite expanse from which the primeval deities Gaia (Earth), Tartarus, Erebus (Darkness), and Nyx (Night) arise. Pythagoras is credited with introducing the term Κόσμος (Cosmos) as the realm of perfect order and arrangement. For Pythagoreans, the imposition of form and limit on the formless and infinite signifies the creation of the universe. They are moral dualists who equate the cosmos with all things "good." An orderly universe must also be beautiful, harmonious, and unified. Chaos thus becomes the repository of all things "bad." The disorderly must be displeasing, dissonant, and strife-ridden. Fundamentally, the cosmos is the realm of being, the presence of "something," while chaos is the realm of dissolution and darkness, or "nothing." Redefining chaos in the Western tradition means challenging the negative values attributed to it as a result of the predominance of dualism and binary logic in Western philosophy.

Postmodernism bears witness to a change in the meaning of chaos. Mathematics, physics (especially research in nonlinear dynamics), information theory, cybernetics, philosophy of science, and—in their own provocative way—literature and literary theory are modes of investigation that have contributed to the new field of chaotics. Hayles remarks that the "science of chaos draws Western assumptions about chaos into question by revealing possibilities that were suppressed when chaos was considered merely as order's opposite" (*Chaos and Order* 3). New definitions of chaos move from the Pythagorean dualism of order and disorder in a more or less fixed opposition to a process of emergence and the interrelation of the two. The new definition of chaos as an unstable and unpredictable relation between order and disorder has been designated by the phrase "orderly disorder."

The shift in the meaning of chaos has acquired, in Kuhnian terms, a claim to the status of a scientific revolution. James Gleick, whose best-selling book, *Chaos: Making a New Science* (1987), provides the first overview of the new field of chaotics, indicates by his subtitle the arrival of a new paradigm. He reports that the "most passionate advocates of the new science go so far as to say that twentieth-century science will be remembered for just three things: relativity, quantum mechanics, and chaos. Chaos, they contend, has become the century's third great revolution in the physical sciences" (5–6). Like relativity and quantum mechanics before it, the science of chaos undermines the tenets of Newtonian physics; it collapses the dream of a world forever governed by deterministic predictability. The theories of Einstein and Bohr, however, pertain to states that are imperceptible in human experience. The weekend golfer strikes a dimpled ball that, never approaching the speed of light, is still effectively subject to Newtonian dynamics. As Gleick observes, "the revolution in chaos applies to the universe we see and touch, to objects at human scale. Everyday experience and real pictures of the world have become legitimate targets for inquiry" (6). Chaotics fulfills Kuhn's criteria for a scientific revolution because it fundamentally changes the apprehension of phenomena as diverse as falling water, the structure of a fern's whorl, the sudden plummet in stock market values, an increase in biodiversity or decline in social orders. In Hayles's estimation, "it is already apparent that chaos theory is part of a paradigm shift of remarkable scope and significance" (*Chaos and Order* 2). Prigogine and Isabelle Stengers make a definitive claim for the "reconceptualization of physics" from deterministic, reversible processes to probabilistic, irreversible ones (*Order Out of Chaos* 232). In their proposition of the emergence of order in thermodynamic systems far from equilibrium, the reinscription of time and the process of becoming—neglected as irrelevancies in classical dynamics—are now essential considerations. There are those, however, who are reluctant to run up the bright flag of revolution. John Horgan, a senior writer at *Scientific American*, chastises science for its alliance with postmodernism and its affair with irony and indeterminacy. The "ironic science," as he calls it, of chaos and complexity is not the rebirth of science but the "end of science" as a pure discipline dedicated to the quest for knowledge and the pursuit of verifiable results (15). In opposition to Lyotard's critique in *The Postmodern Condition* of verification and the special claim to denotation made by scientific discourse (24), Horgan declares that science has entered "an era of diminishing returns" in its definitive knowledge of "physical reality, from the microrealm of quarks and electrons to the macrorealm of planets, stars, and galaxies" and that the few basic forces that govern matter—gravity, electronmagnetism, and the strong and weak nuclear forces—have already been ascertained (16). Sounding as if there were nothing but "normal science" to pursue, Horgan attacks Kuhn's paradigm, Edward Witten's superstring theory, Fred Hoyle's cosmology, Stuart Kauffman's "order for free," and Murray Gell-Mann's complexity theory or "plectics," as desperate forays into unverifiable hypotheses rather than aspects of a scientific revolution.[21] Notably, Horgan's declaration of an approach to the limits of science overlooks Gleick's assertion that chaotics has expanded the realm of scientific inquiry to include the in-between realm of perceptible phenomena that cannot be expressed in purely denotative, deterministic terms; that is, a postmodern science of an essentially ironic universe.

The formation of a new paradigm always introduces a period of disturbance marked by theoretical disagreement, unsettled methodology, and competing claims. Kuhn observes that the invention of alternate theoretical constructions is seldom undertaken by scientists except during the early developmental stages of a new paradigm. The cost of this "retooling" would be deemed too extravagant for the institution or the individual's career at other moments in a science's development (76). The advent of chaotics, however, represents such a period of disturbance in the routine of normal science, and a retooling that introduces competing theories is already evident in the brief history of the field. Hayles calls our attention to two directions of inquiry in chaos theory that emerged by the late 1980s, though both were determined to ignore the findings of the other.[22] The first branch searches for order hidden within chaotic systems. As distinguished from purely random phenomena, this brand of chaos contains deeply encoded structures called "strange attractors." Using computers to process the enormous number of calculations required, a phase space diagram of these chaotic systems reveals that they contract to a confined region and trace complex patterns within it. The second branch is concerned with the order that arises out of chaotic systems. The emphasis here is on "the spontaneous emergence of self-organization from chaos; or, in the parlance of the field, on the dissipative structures that arise in systems far from equilibrium, where entropy production is high" (*Chaos Bound* 9). Like the figure of Chaos in Hesiod's *Theogony*, this model of chaos is seen as the precursor of orderly phenomena rather than its antagonist, instigating the necessarily unstable conditions in which genesis may occur. The strange-attractor branch of chaos theory provides evidence of an immanent design in nature, while the dissipative-structures branch shows the potential for emergent design.[23] In both cases, chaos theory promotes a more complex relation between order and disorder than a binary opposition has allowed. One can now regard the interpenetration of the one through the other with far more clarity than was historically possible. However intellectually convenient it might have been to filtrate forms of order from modes of disorder, chaos theory has shown the dynamic exchange between order and disorder in virtually any open system. Or, in a locution favored by John Barth, order and disorder exhibit a "coaxial esemplasy" of reciprocal influence and transformation (not unlike the cybernetic "feedback loop") (*Further Fridays* 282). There are, naturally, differences in the retooling necessary for the study of chaos. Hayles remarks that the "strange-attractor branch differs from the order-out-of-chaos paradigm in its attention to systems that remain chaotic," while the latter approach favors chaotic systems that generate orderly behavior. The immanent-design branch focuses on "the orderly descent into chaos rather than on the organized structures that emerge from chaos" (*Chaos Bound* 10). As William Carlos Williams says, "The descent beckons / as the ascent beckoned" (*Collected Poems* 245), and the relationship studied by chaotics is a reciprocal or coaxial one.

As the chief chronicler of the first branch of chaos theory in *Chaos*, Gleick recounts the research of Edward Lorenz in the "butterfly effect," David Ruelle in turbulence, Mitchell Feigenbaum in universal behavior, Benoit Mandelbrot in fractal geometry, and Robert Shaw in "strange attractors." Lorenz was fascinated with the weather as a system "displaying familiar patterns over time. . . . But the repetitions were never quite exact. There was pattern, with disturbances. An orderly disorder" (15). In the linear dynamics of Newtonian physics, small causes lead to small effects; big causes have big effects. But Lorenz realized that in the weather, functioning as a nonlinear system, small causes could be amplified to create large-scale effects. Thus he coins the "butterfly effect" that not so fancifully proposes that an insect flapping its wings in Baja California could cause a tornado in South Florida. The "butterfly effect," or sensitive dependence on initial conditions, accounts for the rich turbulence—the unpredictability—of weather at the same time that certain recursive—though never exactly repetitious—patterns emerge. In postmodern literature one observes a pervasive concern with randomness and the condition of uncertainty. DeLillo's protagonist in *White Noise*, Jack Gladney, expresses his exasperation at his inability to state the probability of rain in a precise locality: it's a "victory for uncertainty, randomness, and chaos. Science's finest hour" (24).

Benoit Mandelbrot's dissatisfaction with Euclidean geometry because it "failed to capture the essence of irregular shapes" (Gleick 96) led to his proposal of fractional dimensionality. As Gleick explains, "Fractional dimension becomes a way of measuring qualities that otherwise have no clear definition: the degree of roughness or brokenness or irregularity of an object. A twisting coastline, for example, despite its immeasurability in terms of *length*, nevertheless has a characteristic degree of roughness." Mandelbrot's claim was that "the degree of irregularity remains constant over different scales. . . . Over and over again, the world displays a regular irregularity" (98). Thus a second principle of chaotics is self-similarity across scales, or as Hayles puts it, a "recursive symmetry. A figure or system displays recursive symmetry when the same general form is repeated across many different length scales, as though the form were being progressively enlarged or diminished" (*Chaos and Order* 10). Mandelbrot's description of the fractal curve in coastlines, a conch shell, crystals, and snowflakes "implies an organizing structure that lies hidden among the hideous complication of such shapes" (Gleick 114).

One of the most intriguing figures of immanent design in chaos theory is the "strange attractor." Physicists were familiar with simple kinds of attractors, the point of a system's cycle to which the system is attracted. Fixed-point attractors, such as the midpoint of a pendulum's path, represent behavior that reaches a steady state; limit cycles represent behavior that repeats itself continuously. For complex, nonlinear systems, Lorenz, Shaw, and others discovered the existence of "strange attractors." Hayles explains that these "complex systems follow predictable paths to randomness and trace recognizable patterns when they are mapped into time-series diagrams" (*Chaos and Order* 8) (Figure 1). Although any particular orbit in a nonlinear system diverges unpredictably from its predecessor, over many thousands of orbits, the points in the phase space diagram "continue to evolve within a confined region." The "strangeness" of the strange attractor is that it "combines pattern with unpredictability, confinement with orbits that never repeat themselves" (*Chaos and Order* 9). This is another example of the presence of "regular irregularity," or orderly disorder. Gleick assesses Lorenz's discovery of a strange attractor in fluid convection, concluding that "Nature was *constrained*. Disorder was channeled, it seemed, into patterns with some common underlying theme. . . . An attractor like Lorenz's illustrated the stability and the hidden structure of a system that otherwise seemed patternless" (152–53).

The phrase "order hidden in chaos" somewhat misrepresents the relation between order and disorder described by the scientists in Gleick's book. The patterns "revealed" by Lorenz, Mandelbrot, or Shaw in dauntingly complex systems were never really "hidden." Their presence is only now ascertainable through the new science of chaos. The misapprehension would be to assume that we can now perceive more of order than we were previously able; or that the uncovering of some order where only disorder had been found was the sole endeavor. In fact, the presence of "regular irregularity" in nonlinear systems suggests a different condition that is neither the exclusive domain of orderliness or disorderliness. Fractal symmetry and strange attractors suggest that an "inherent opposition" between regularity and irregularity, order and disorder, was a limiting myth that sustained human inquiry for some time but now must be superseded. I prefer to follow Michel Serres's terminology of "interpenetration" because it refutes the tendency to isolate the patterned, the solvable, and the elegant simplicity against the mottled, the unsolvable, and the fearsomely complex. He rejects the "dualist hell" of reason or irrationality, science or religion, unity or multiplicity, noise or harmony. For Serres, "What fluctuates is order and disorder. What fluctuates is their vicinity and common border, their

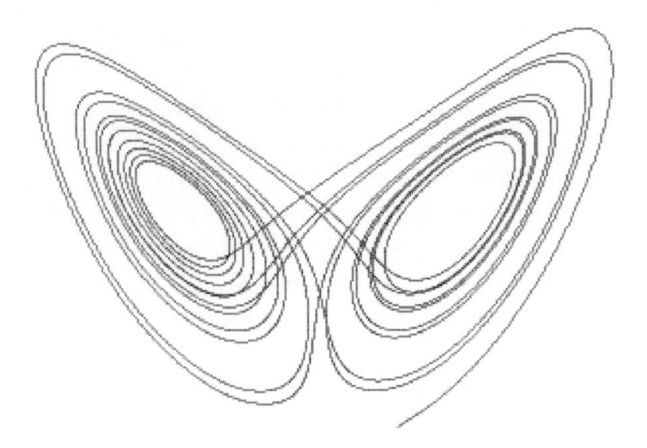

Figure 1. The Lorenz Attractor

relationship and mutual penetration" ("Dream" 233). He proposes that the "universe vibrates endlessly between" a unitary system and an incomplete multiplicity, that we will find pattern revealed where chaos is apparent, and disorder present where the normalizing eye or brain had found regularity.

The second branch of chaos theory focuses on the order that arises out of chaotic systems. In *Order Out of Chaos: Man's New Dialogue with Nature*, Prigogine and Stengers argue that in conditions far from equilibrium "order and organization can actually arise 'spontaneously' out of disorder and chaos through a process of 'self-organization'" (xv). The paradigm shift here is quite apparent, and remains controversial in scientific communities, including those represented by Gleick. Chaos and order, randomness and structure, had been perceived as inimical, antagonistic conditions: chaos destroys order where it encounters it; order needed to be asserted against chaos in return. But Prigogine claims that these highly entropic, "dissipative systems" are capable of creating pockets of higher organization, complexity, or negentropic islands without violating the second law of thermodynamics (namely, that in a closed system entropy cannot decrease for any spontaneous process). In effect the far-from-equilibrium system serves as an incubator for new expressions of order. Prigogine's hypothesis, based in part on self-organizing or "feedback" chemical reactions, attempts to resolve a dispute between the biological sciences, in which more complex environments and species evolve through apparently random interactions, and thermodynamics, whose laws presage the inevitable "heat death" of a universe gradually running down to absolute zero. Prigogine and Stengers ask "why complexity seems spontaneously to come into existence." They envision "a world that can renew itself rather than a universe that is constantly running down" (Hayles, *Chaos and Order* 12). In arguing against "heat death," Prigogine and Stengers "envision entropy as an engine driving the world toward increasing complexity rather than toward death. They calculate that in systems far from equilibrium, entropy production is so high that local decreases in entropy can take place without violating the second law. Under certain circumstances, this mechanism allows a system to engage in spontaneous self-organization" (Hayles, *Chaos and Order* 13).

Prigogine is very much aware of the cultural and philosophical implications of his research, arguing that the types of interactions that occur at the chemical and biological level also appear at the social levels of evolution, albeit on a much shorter time scale. His work is directed toward the human universe of open systems, fluctuations, and irreversible time. As Alvin Toffler remarks:

> while some parts of the universe may operate like machines, these are closed systems, and closed systems, at best, form only a small part of the physical universe. Most phenomena of interest to us are, in fact, *open* systems, exchanging energy or matter (and, one might add, information) with their environment. . . .
>
> In Prigoginian terms, all systems contain subsystems, which are continually "fluctuating." At times, a single fluctuation or a combination of them may become so powerful, as a result of positive feedback, that it shatters the preexisting organization. At this revolutionary moment—the authors call it a "singular moment" or a "bifurcation point"—it is inherently impossible to determine in advance which direction change will take: whether the system will disintegrate into "chaos" or leap to a new, more differentiated, higher level of "order" or organization (xv).

Prigoginian cosmology seems to present an ambivalent message. In positive terms his world does not cede to the inevitable decay into homogeneity and the erosion of functionality that is attendant on entropy. The possibility of negentropic pockets of increasing differentiation and higher levels of order arising tends to be reassuring to beings such as ourselves; such a possibility at least explains our presence in the thermodynamic cloud of energy in space. The emphasis on *self*-organization in the Prigoginian cosmos makes the intervention of a god in creation unnecessary. A sustained complexity also suggests that the universe need end neither in catastrophe nor in exhaustion, neither with a bang nor a whimper. The negative import resides in the inherently unpredictable nature of dissipative systems. Order is not the "outcome of some preconceived plan" ("Order Out of Chaos" 41). In states of great instability and fluctuation, order may arise and/or structures may be destroyed or reorganized—but such outcomes would require an impossibly precise observation of initial conditions to predict. If anything, it's a more uncertain universe than that offered to us by the second law of thermodynamics.

Both approaches to chaos—whether they propose an immanent or an emergent design in chaos—focus on the complicated relation of order and disorder rather than their antithesis. Chaos has now been defined as a condition that lies between a state of complete determinism and a state of utter randomness. This division—however complementary—of chaos theory into two branches has been unsettled somewhat by recent studies in the science of complexity, especially at the Santa Fe Institute founded by George Cowan in the mid-1980s. In his narrative of the birth of complexity studies, Mitchell Waldrop situates self-organizing complex systems at a balancing point between order and chaos; that is, between static objects such as the snowflake or computer chip that are merely complicated and the unpredictable behavior of chaos (12). The spontaneous, adaptive systems that describe complexity reside in a narrow border "on the edge of chaos." And yet it is in this corridor between order and disorder that habitable environments, sophisticated social systems, and the active exchange of information that sustains them, may arise. An alternative opinion advanced by Chris Langton situates the chaos described by nonlinear dynamical systems, such as strange attractors, in which sometimes simple deterministic equations produce tremendously divergent behavior, as a subset or special case of complexity (Lewin 12). Furthermore, Stuart Kauffman suggests that some nonlinear systems in nature—such as an eddy swirling in a swift-moving stream—diverge dramatically over time (one definition of chaotic behavior); but many other complex systems, perhaps the larger part, "produce convergent flow, produce structure. This applies in our evolutionary models in computers and in biological systems" (Lewin 186). (See Table 1, which shows the relationship of order and disorder in the emerging science of chaoplexity.)

James Yorke, credited with having first used the term in its current usage in "Period Three Implies Chaos" (1975), has said that "chaos refers to a restricted set of phenomena that evolve in predictably unpredictable ways—demonstrating sensitivity to initial conditions, aperiodic behavior, the recurrence of certain patterns at different spatial and temporal scales" (Horgan 196). Chaotic systems would be limited to those for which simple sets of deterministic mathematical rules, when computed through many iterations, give rise to unpredictable patterns that never quite repeat themselves. From the perspective of complexity theorists, these chaotic systems (Mandelbrot's fractals, strange attractors), fail to display the adaptive behavior for which

Table 1. From Order to Disorder

Order	Complicated Structure	Complex Adaptive Systems	Chaos	Disorder
Salt crystals; algebraic notation of chess.	Snowflakes; high-rise office building; computer chip.	The brain; the stock market; speciation.	Turbulent flow; fractals; the butterfly effect.	Random number generation; random motion of gas molecules.

complex systems—especially living organisms—are noted. In rejecting the possibility of a single theory that would embrace all complex phenomena, Murray Gell-Mann still prefers his neologism, "plectics," to encompass the "relation between the simple and the complex, and in particular how we get from the simple fundamental laws that govern the behavior of all matter to the complex fabric that we see around us" (Horgan 213). Some regard complexity as the result of order so highly patterned that the processes of determination that led to its current condition are no longer perceptible or capable of being resolved by any simple procedure. It is as if the complex system has "forgotten" its initial conditions. Gell-Mann proposes, however, that complexity can be generated from simple equations. He calls this "surface complexity arising out of deep simplicity" (Lewin 14). The complexity of these systems is defined by their adaptivity, as they "encrypt information about their environment" (15), seeking out and incorporating pattern. By contrast, Prigogine's dissipative structures rely on instability as their operative process in the potential production of increased organization. To all but the initiate the complex state and the chaotic state may appear formally identical, whether that condition was arrived at by the adaptivity or the instability of the system.[24]

Perhaps the broadest appeal of chaos and complexity theory is the application of these methods of inquiry to the "middle distance," or phenomena that are directly accessible to our perception. Human activity in the middle dimension has seemed to conduct itself by rules other than those that governed the microscopic structure of elementary particles studied in quantum physics or the macroscopic structure of the universe examined by cosmology. Many of the open systems encountered in chaos and complexity theory are readily apparent components of the human environment: the unpredictable and yet recognizably patterned weather systems; the shift from laminar to turbulent flow in smoke rising or water flowing downstream; the rise and sudden decline in the population of a species; the rise and fall of a certain commodity's price on the stock market, or of the market as a whole; or the evolution of an ecosystem. Because these are open systems exchanging matter, energy, or information with their environment, their investigation has particular resonance with human activity as studied in economics, politics, anthropology, and of course, literature. Interdisciplinarity is an almost inevitable consequence of chaos and complexity theory as open dynamical systems in different fields respond to analysis. As Prigogine remarks (albeit with a masculine bias), "We cannot define man in isolation; his behavior depends on the structure of the society of which he is a member, and vice versa. This structure evolves as a result of the action taken by individuals. It is perhaps in human society that this interaction between units and global structure is the clearest" ("Order Out of Chaos" 42). Chris Langton, one of the proponents of complexity theory, also points to the interaction between "emergent global structure" and "local interaction." The global structure (of whatever scale) seems to have stable and consistent properties; the local components appear to behave randomly. But "from the interaction of the individual components *down here* emerges some kind of global property *up here,* something you couldn't have predicted from what you know of the component parts. . . . And the global property, this emergent behavior, feeds back to influence the behavior of the individuals *down here* that produced it" (Lewin 12–13; Langton's emphasis). Such order arising out of complex dynamical systems has a holistic quality to it; it might even be called a New Age science. The theory presumes that there are simple principles that underlie the most complex systems, and that these principles pertain regardless of the scale or the nature of the components in the system. The study of bifurcations and fluctuations, period doublings and limit cycles, self-similarity and adaptivity, may not offer answers to all problems of the physical universe or the human condition. But the postmodern age seems especially concerned with the increasing complexity revealed in the environment in which we exist, in the systems that we've created to navigate that environment, and in the relationships with which we coexist.

As a discipline literature has been examining the complexity of the human dimension for some time. Literature has, arguably, expended far more of its interpretive and analytical abilities on the intrusion of chaos in human lives than it has in the assertion of whatever order might be found or created there. Art has always regarded the human condition as an open system. Recently, however, there has been a more pronounced convergence of the attentions of literature and science. As I maintain, postmodern fiction can be characterized by its attention to the interrelation of order and disorder. The world-making of postmodern fiction sometimes anticipates and sometimes confirms the scientific theories of chaos and complexity. Methodologies such as the self-reflexivity of literary forms and the recursive symmetry of fractal geometry, nonlinear narrative and indeterminacy, generative procedures and self-organization, suggest an inevitable correlation between the two disciplines. This convergence of inquiry into the relation of design and debris, order and disorder, system and distribution, is so essential to the present condition of knowledge that it can be found in the several discourses of philosophy, physics, mathematics, literary theory, and fiction.

Proceduralists and Disruptors

Let me propose a correspondence between the methodology of postmodern fiction and the two directions of inquiry in chaos theory. Works such as Barth's *LETTERS* (1979), Christine Brooke-Rose's *Amalgamemnon* (1984), Italo Calvino's *The Castle of Crossed Destinies* (1976), Harry Mathews's *The Journalist* (1994), Milorad Pavic's *Dictionary of the Khazars: A Lexicon Novel* (1989), Georges Perec's *Life, A User's Manual* (1978), and Sorrentino's *Misterioso* (1989) express an immanent design that is revealed deep within the chaos of their materials. These writers can be designated as "proceduralists." They formulate a plan comprised of arbitrary and exacting rules, carrying it out in spite of—or in anticipation of—the narrative consequences. The procedural form of their work consists of predetermined constraints that are relied upon to generate the content, trajectory, and orderliness of the work during composition. Brooke-Rose, for example, restricts the composition of her novel to the non-realized verb tenses (future, subjunctive, or interrogatory) whose subjective inquiry and speculative vision govern the mood of the narrator and her prospects. Barth traces the design of his epistolary novel, *LETTERS*, on the calendar for the months of March through September, 1969, setting the order of exchange between the seven correspondents (including the author "himself"), and recapitulating characters and plots from his six previously published volumes of fiction. In his foreword to a new edition of the novel, Barth remarks, "Complex? Well, yes. Complicated? For sure. Designed and constructed with a certain rigor? You bet." In his own considered opinion the novel is "enormously readable . . . *despite* its elegant construction" (*LETTERS* xvii).

The Paris-based Oulipo (or Workshop of Potential Literature), whose members include Calvino, Mathews, Perec, and Raymond Queneau, favors the use of mathematical formulae in literary production. One of the masterpieces of the genre, nearly as capacious as Barth's novel and more elaborate in its design, is Perec's *Life, A User's Manual*. The eccentric English millionaire Percival Bartlebooth commits himself to a fifty-year regimen in which he creates 500 watercolors of seaports in an identical format (65 cm × 50 cm); they are shipped back to Paris where the puzzle-master, Gaspard Winckler, converts them into jigsaw puzzles of 750 pieces; after the puzzles have been solved in order, the pictures are removed from their backing and dissolved. Bartlebooth's resolution, that "his whole life would be organised around a single project, an arbitrarily constrained programme with no purpose outside its own completion" (117), stands as an axiom of Oulipian artistry. The novel itself is a series of ninety-nine interlocking stories (and an epilogue) that are arranged with reference to the residents of an apartment building at 11, rue Simon-Crubellier. Movement from one apartment to another in the novel is based on the knight's tour problem in chess; that is, the path the knight would have to follow so as to travel all the 8 × 8 squares of the chessboard touching each square only once. Perec extends the grid to 10 × 10 squares, but there is a "deliberate imperfection," a Lucretian *clinamen:* the bottom left-hand square (1,0) is passed over between chapters 65 and 66.[25] That error is the sign of the artist. These self-imposed constraints of structure or language have a paradoxically liberating and productive effect; rather than inhibit the production of the text, they actually can be said to enable and enhance its making. The work becomes a island of negentropy in a sea of disorderliness. Far from seeing it as a reductive product of its rules, Perec regards the art object—watercolor, jigsaw puzzle, or novel—not as "the sum of elements to be distinguished from each other and analysed discretely, but a pattern, that is to say a form, a structure: the element's existence does not precede the existence of the whole, it comes neither before nor after it, for the parts do not determine the pattern, but the pattern determines the parts: knowledge of the pattern and of its laws, of the set and its structure, could not possibly be derived from discrete knowledge of the elements that compose it" (*Life* 189).

Postmodern proceduralism bears comparison to the strange-attractor branch of chaotics. Often simple, deterministic rules can produce complex results in a nonlinear system, as these narratives demonstrate. The presence of an arbitrary constraint presumes that complex patterns will become apparent, and that the very constructedness of the text will take shape of its own accord. The text is neither an incidental repetition of some permanent or transcendent order in the universe nor simply an individualistic expression of the author's own determination. The author may set the rules that generate the text, but he cannot fix the patterns that result. In his essay on the affinities of postmodernism, chaos theory, and arabesque design, Barth points out that these complex systems share feedback mechanisms "in which output loops back into the system as input" (*Further Fridays* 333). The patterning that emerges—in the Mandelbrot set, Persian carpets, or Barth's arabesque novels—is the product of rules that govern the system, but the patterning is not predictable from the rules alone. During each iteration of the design, output is continually channeled back into the system as input, changing the system for subsequent iterations. "Another way to put it (and to novelize it, as I attempted to do in *The Last Voyage of Somebody the Sailor* [1991]) is that the voyager is changed by the voyage, but the voyage is also changed by the voyager: dynamic feedback loops. Coaxial esemplasy" (*Further Fridays* 334). There are many potential combinations and structures in a dynamical system, but feedback makes it impossible to predict which will be asserted and which not. The order is in the making. Barth concludes that "both varieties of chaology are of potential literary interest. The order-out-of-chaos approach, for instance, obviously applies to the way that poems and novels . . . come into existence. But it is the strange-attractor branch of chaology that strikes me as having more affinity with Postmodernism and the arabesque" (*Further Fridays* 335). These are the figures of an immanent design.

Distinct from the texts of postmodern proceduralism that manifest a rigorous plan are those that fiercely avoid the imposition of a determining structure. Works such as Acker's *Blood and Guts in High School* (1978), Paul Auster's *The Music of Chance* (1990), William Burroughs's *The Ticket that Exploded* (1967), William Gaddis's *A Frolic of His Own* (1994), Joseph McElroy's *Lookout Cartridge* (1974), and David Foster Wallace's *Infinite Jest* (1996) propose that out of the vortex of their disorderliness a capacity for self-organization may emerge. These writers can be designated as "disruptors." They surrender themselves willingly to, and just as often exploit, the inevitable debris of culture and decay of systems in narratives capable of incorporating the wholly unpredictable. They disdain the readerly conveniences of a linear narrative, with its neat proportionality of cause and effect, which is deemed inadequate to the complexity of the postmodern condition. These are works that exhibit a great deal of surface complexity, and only reveal their principles of organization after much scrutiny. Two characteristics that they hold in common are a cynicism regarding the thin patina of order that is promoted by hegemonic cultural institutions to deliver the illusion of a far greater stability than they actually maintain; and a profound paranoia regarding the origin and design of power structures not readily apparent, but which may be exerting a fundamental directive force in human affairs or the physical universe. The title of

Auster's *The Music of Chance* suggests how "random, accidental encounters" (1) might still enable one to understand one's "place in the invisible order of things" (10). His nonfiction *The Red Notebook* likewise expresses a fascination with serendipity. These are not instances of all-too-convenient coincidence, which are ultimately meaningless; but rather of serendipity, the loss of control, the absence of design, that nevertheless may reveal meaning. Recoiling from the causality and rationality that bolster conventional narrative, Auster turns to "the presence of the unpredictable, the utterly bewildering nature of human experience. From one moment to the next, anything can happen. Our lifelong certainties about the world can be demolished in a single second. In philosophical terms, I'm talking about the powers of contingency" (*Art of Hunger* 278–79).

Burroughs and Acker employ the aleatory technique of the cut-up as a deliberate disruption of received associations and complacent forms of order. These structures have been crafted by the state in the sole interest of furthering its consolidation of power and control over the individual. In his epilogue to *The Ticket that Exploded,* Burroughs chides the "invisible generation" of conformists that "the psychological liberation achieved as word lines of controlled association are cut will make you more efficient in reaching your objectives" (208). In fact these random processes are nothing like a model for efficiency in the capitalist, bureaucratic, and heterosexual system that Burroughs assaults. But he insists that "the first step is to isolate and cut association lines of the control machine." Only in cutting and splicing the "old prerecordings" (217) is the individual liberated from the straitening control of the state. Acker emphasizes the destructive and destabilizing force of desire in her crusade to thwart the repressive hegemony of the patriarchy. Here I reproduce her "shouted" capitals from a passage in *Blood and Guts in High School,* as one example of her resistance to the regularization of thought through the use of spatial form, typography, and images in the text: "EVERY POSITION OF DESIRE, NO MATTER HOW SMALL, IS CAPABLE OF PUTTING TO QUESTION THE ESTABLISHED ORDER OF A SOCIETY; NOT THAT DESIRE IS ASOCIAL; ON THE CONTRARY. BUT IT IS EXPLOSIVE; THERE IS NO DESIRING-MACHINE CAPABLE OF BEING ASSEMBLED WITHOUT DEMOLISHING ENTIRE SOCIAL SECTIONS" (125). Burroughs and Acker are prominent examples of a literary terrorism that attacks the sociopolitical hegemony of Western institutions. But lest a spiral descent into the maelstrom seem the only path they might take, Burroughs's "counterrecording and playback" and Acker's self-determination through desire suggest that a new construction of order may arise after the limiting and oppressive strictures of an established order have been destroyed.

After suggesting that entropy is a necessary and unavoidable process in business and financial systems in his novel *J R* (1975), Gaddis turns, in *A Frolic of His Own*, to the unextirpated presence of disorder in the very system whose purpose is the establishment of immediate cause and liability: the law. Rather than lend itself to a celebration of mastery, as one would expect of a "systems" novel,[26] *Frolic* is a spectacle of impotence and disability, as the legal system fails to impose order on the disorderly conduct of its subjects. The narrator, in one of his few interjections in the novel, consigns the litigious Oscar Crease to "the historic embrace of the civil law in its majestic effort to impose order upon? or is it rather to rescue order from the demeaning chaos of everyday life" (28). In aspiring to an all-encompassing and deterministic system, jurisprudence fails to withstand the irruption of contingency, the perfidy, and the nagging irresolution in human affairs. It fails as "a vehicle for imposing order on the unruly universe" (459). Natural law remains always outside of the governance of the juridical law. The legal system's sole recourse to language as the means to its construction is undermined by the instability of the signifier. Gaddis's dialogic style, with its run-on syntax, unstable punctuation, and abrupt transitions between speakers emulates a world in which flux and mutation outstrip the best efforts of deterministic control.

As chaos is a precursor of any orderly structure that arises in these novels, the postmodern fiction of disruption is more closely related to the order-out-of-chaos approach to nonlinear dynamics. The texts of Acker, Burroughs, Gaddis, McElroy, Pynchon, and Foster Wallace should be treated as open systems that require the input of energy and information from the reader to be fully realized as works of art. Umberto Eco describes such plural, incomplete, and unstable but generative texts as *opera aperta*, open works. Their "tendency toward disorder," he remarks, "characteristic of the poetics of openness, must be understood as a tendency toward *controlled* disorder, toward a circumscribed *potential*, toward a freedom that is constantly curtailed by *the germ of formativity* present in any form that wants to remain open to the free choice of the addressee" (*Open Work* 64–65; Eco's emphasis). Eco describes the delicate oscillation between disorder and organization in this class of texts. Their tendency toward disorder encourages the reader to make performative choices in the text; that is, choices that will constitute both the meaning and structure of the text. Without such performativity, the work would drift into an undifferentiated formlessness. The disorderly openness of the text nevertheless retains the potential for demonstrating new order. Ilya Prigogine refers to the presence of such controlled disorder in open systems as "dissipative structures." Systems far from equilibrium that interact with their surroundings are characterized by high entropy production. Dissipative structures require more energy to sustain them than the stable, repetitive forms that are incapable of further development. In their interaction with their surroundings these dissipative structures facilitate self-organization. The postmodern fiction of disruption is likewise characterized by its openness, instability, and high entropy production. Its complex textuality demands interaction with the reader—often at the cost of great effort, which some readers are unwilling to expend—at every moment. But it is precisely because of the continuous struggle and exchange that disruptive fiction is capable of self-organization, or as Eco puts it, possesses the germ of formativity. These are the fictions of an emergent design.

These categories of procedural and disruptive fiction are naturally subject to an inherent instability, as are all phenomena described by chaotics. The fiction of immanent design such as Hawkes's *Travesty*, Barth's *Last Voyage of Somebody the Sailor*, and Coover's *Universal Baseball Association* that is discussed in the following chapters merges with rather than departs from the fiction of emergent design, represented by Acker's *Empire of the Senseless*, DeLillo's *White Noise*, and Pynchon's *Gravity's Rainbow*. Gilles Deleuze and Félix Guattari speak of the continual reterritorialization of striated (restrictive, orderly) and smooth (generative, disorderly) space. In their distinction between a sedentary space and a nomad space, they acknowledge that "no sooner do we note a simple opposition between the two kinds of space than we must indicate a much more complex difference by virtue of which the successive terms of the oppositions fail to coincide entirely. And no sooner have we done that than we must remind ourselves that the two spaces in fact exist only in mixture: smooth space is constantly being translated, transversed into a striated space; striated space is constantly being reversed, returned to a smooth space" (474). Just as order and disorder are discovered to be closely interrelated, so the methodologies of the proceduralists and the disruptors may appear to be in continual exchange and express themselves simultaneously in any given work. Postmodern fiction illustrates a complicated blending of system and distribution, formalism and flux, procedure and process, in its composition. Any literary model is preeminently a space in which a mixture of the smooth and the striated, the orderly and the disorderly, will be present, rather than the exclusive dominance of the one over the other. So postmodern fictions perform an elegant and ongoing metamorphosis between figures of design and debris.

2

Design and Debris

John Hawkes's *Travesty,* Chaos Theory, and the Swerve

> To open the route, way, track, path in this incoherent chaos, this tattered cloud, whose dichotomic thicket is reformulated in the common space of transport when it is reconstructed. To find the relation, the logos of analogy, the chain of mediations, the common measure, the asses' bridge; to find the equilibrium or the *clinamen*.
>
> —Michel Serres, *Hermes*

John Hawkes's short novel, *Travesty* (1976), lends itself to synopsis in much the same way as an elegant experiment in physics clearly and precisely defines its parameters and hoped-for results. "In the darkest quarter of the night" (11), an elegant and finely tuned performance car speeds down a country road in the south of France at nearly one hundred and fifty kilometers per hour. The unnamed driver expertly guides his machine as he explains, in an extended and unbroken monologue, to his friend the poet, Henri, sitting beside him, and to his daughter, Chantal, uncomfortably crouched in the back seat, that he intends to end their journey quite deliberately with a swerve into the thick stone wall of a desolate farmhouse. Their perilous route and destination have been carefully plotted, and the driver estimates their arrival in one hundred minutes "by the dashboard clock" (25). The driver reveals that he knows that Henri has taken both his wife, Honorine, and Chantal as his mistresses. The passengers are left little room to maneuver: the driver deplores Chantal's fits of vomiting and hysteria, and blandly points out that Henri's intervention at the wheel will "pitch us into the toneless world of highway tragedy even more quickly than I have planned" (11). Although he remains undissuaded by Henri's attempts at rational argument, the narrator encourages in the time remaining an examination of the perfect "symmetry" (25) that their mutual destruction will create. The subjects of his experiment are thus engaged in its analysis for the duration of the novel.

Travesty is already in motion when it is first encountered, and it remains constantly in motion for its brief duration. The driver has achieved his precipitous acceleration to a surely fatal speed. The trap has been sprung on his hapless victims. In this sense the cause of the present action is matter for retrospection only, seen dimly receding into the night. The initial conditions —infidelity, betrayal, the artistic temperament—are by no means precisely determined and remain shrouded in uncertainty. The reader in turn becomes a victim of the driver's monologue, another passenger whose pleading remains unheeded and unrecorded throughout the terrifying ride. Hawkes most resembles Edgar Allan Poe in his allegiance to calculated trauma, undisturbed terror, and a dark futility that will not allow the reader to suspect that Henri could at last convince the driver to relent.[1] So despite the linear progress of the novel, the narration refutes any claim to suspense as a means of galvanizing the reader's attention. Hawkes bypasses the pedestrian concern with *what* will happen to the characters assembled and proceeds to a discussion of *why* they are thus subjected to extremes. Reminiscent of Poe's philosophy of composition, the narrator wishes to create a single effect upon his passengers:

> We have agreed on the surface aspects of trauma: the difficulty of submission, the problem of surprise, a concept of existence so suddenly constricted that one feels like a goldfish crazed and yet at the same time quite paralyzed in his bowl. A mere question of adjustment. But the fact of the matter is that you do not share my interest in what I have called "design and debris." (19)

The driver would have his captive audience believe that the collision is unavoidable, and that their time in his presence is best spent studying the procedures and probable outcome of his experiment. He refutes the charge that he is "merely some sort of suicidal maniac, an aesthetician of death at high speed" (18). He preys upon the normal human instincts to avoid accident and to fear death, but he is not chiefly concerned with inflicting punishment on those who have betrayed him. That would be petty emotionalism, gross morbidness. He begs Henri to adjust his abhorrence of vehicular collision in order to appreciate his own "reverential amazement" before "the symmetry of the two or even more machines whose crashing" is all too swiftly erased by the authorities (20). In an anecdote related to the composition of *Travesty*, Hawkes relates a fascination with vehicular mayhem quite comparable to that of the driver: "Two years ago this summer there was a car accident near us in Brittany—a marvelous French accident with cars coming together head-on and then just melding their pieces all over the landscape for hundreds and hundreds of yards" ("A Conversation" 165). Two highly organized machines disintegrate, disrupting the orderly flow of traffic; the "melding" of the automobiles forms a new and equally complex (dis)array, subject to the aesthetic consideration of the properly attuned observer. In the confusion and disorder of such a site, the driver views a "symmetry" revealed. Like Hawkes, the driver is an aesthetician of collision; he admires the scattering of debris and does not fear it. In a poetics of postmodernism, he worships before dispersal.[2]

Unimpeded flow, deflection, collision: for the duration of the novel the sports car is a particle in movement on a preordained path. But the final swerve and impact of the car into the farmhouse wall are beyond the bounds of narratorial possibility. The driver makes a promise before he ceases to speak, and hence ceases to be recorded: "there shall be no survivors. None" (128). The driver is an exemplar of the unreliable narrator, especially of the sort one encounters in French New Novels such as Alain Robbe-Grillet's *The Voyeur* (1955) or *Jealousy* (1957). It is possible that this scenario is entirely concocted within the obsessed mind of the cuckolded husband and betrayed father as he sits in an armchair at his mansion, rather suspiciously called "Tara." There are probably as many scenarios as there are clever critics that call into question the palpable reality of the events described in the novel. Hawkes states, however:

> I don't want to be left with nothing but the narrator. . . . Some reviewers haven't wanted to admit that there is a car accident, either. I think the accident the narrator imagines is the accident that occurs. Without the literal accident, you wouldn't have the impossible object, you wouldn't have the whole fabric of imagined event, you wouldn't have the imagination exemplified as it is in that short novel. ("A Conversation" 169–70)

Hawkes is repelled by the suggestion that his narrator merely indulges a morbid fantasy. The driver's act of imagination, his aesthetic impulse, must inevitably come into contact with the cold stone wall of reality, the literal accident. The crash thus articulates the aesthetic of "design and debris" beyond theory or simulation.[3] Art happens not in the evasion of, but in the confrontation with, the real. Nevertheless, it remains impossible for this novel to achieve the totality of closure: the impact and its aftermath can only be imagined, never described, by the narrator. The three characters must always be suspended in motion as they race toward, but never attain, their final destination. The narrative is thus a self-consuming, self-destructive artifact. If, as Hawkes argues, we are to accept the narration as a literal description of events as they transpire, how or by whom is this monologue recorded? The "impossible object" that Hawkes describes is both the collision and the novel itself. The sheer impossibility of the narrative moment (there can be no preterit recollection) flaunts the very artifice of all fiction; all art is an impossible object. And for the driver, his gloved hands tightly gripping the wheel, the aesthetic object that is the debris of collision can never be experienced. Both the collision and *Travesty* itself are painstakingly literal and blatantly artificial.

The Perfectly Contrived Accident

Because the driver cannot experience the product of his collision, his attention focuses on the present process by which he attains his imagined end. Henri views himself as the unfortunate dummy in a crash-test that his friend has set into motion; he cannot remove his attention from the farmhouse wall. Meanwhile, the driver advances his "theory" of art, that "ours is the power to invent the very world we are quitting. . . . It is as if the bird could die in flight. And unless we exercise this power of ours we merely slide toward the pit feet first, eyes closed, slack, and smiling in our pathetic submission to an oblivion we still hope to understand" (57). The driver refutes the fondly held perception of life as a slow, undisturbed, laminar flow toward an emission that is beyond comprehension. He welcomes turbulence as an intrinsic factor of art and existence alike. In *Chaos: Making a New Science*, James Gleick concurs that "chaos is a science of process rather than state, of becoming rather than being" (5). The driver fails to be excited by the laminar flow of traffic on a limited-access interstate, his cruise-control engaged for hundreds of miles, nor is he particularly engaged by the car safely parked in its garage. Like the bird that dies in flight, it is the instantaneous phase transition from laminar to turbulent flow that appeals to him. Gleick describes the physical manifestations of this aesthetic and scientific mystery:

> When flow is smooth, or laminar, small disturbances die out. But past the onset of turbulence, disturbances grow catastrophically. This onset —this transition—became a critical mystery in science. The channel below a rock in a stream becomes a whirling vortex that grows, splits off and spins downstream. A plume of cigarette smoke rises smoothly from an ashtray, accelerating until it passes a critical velocity and splinters into wild eddies. (122)

The onset of turbulence, that moment in which the orderly flow of molecules becomes random, can be "seen and measured in laboratory experiments . . .

but its nature remains elusive" (122–23). For Hawkes, the imaginative act resembles this phase transition between invention and destruction, the calculable and the inscrutable, the process of life and the stasis of death.

Regrettably, the driver will be unable to measure the phase change of his automobile from laminar to turbulent flow, from order and coherence to disorder and fragmentation, but he does provide a series of explicitly theoretical propositions while the car traverses the French countryside. Focusing on the didactic rather than on the traumatic, visceral power that these passages possess accedes to the driver's desire to convince his passengers (and thus the implied reader) that the success of his aesthetic experiment is worth the cost of their lives. But his propositions are so intriguing that they merit attention even in isolation from the thrust of the narrative. In the earliest enunciation of the driver's theory, he appears to sympathize with Henri's incomprehension:

> It is not easy to discover that your closest friend and husband to one mistress and father of the other is driving at something greater than his customary speed, at a speed that begins to frighten you, and that this same friend is driving by plan, intentionally, and refuses to listen to what for you is reason. What can you do? How in but a few minutes can you adjust yourself successfully to what for me is second nature: a nearly phobic yearning for the truest paradox, a thirst to lie at the center of this paradigm: one moment the car in perfect condition, without so much as a scratch on its curving surface, the next moment impact, sheer impact. Total destruction. In its own way it is a form of ecstasy, this utter harmony between design and debris. (17)

The driver challenges Henri's notion of what is reasonable or rational, since such humanistically defined conceptions contribute directly to Henri's inability to appreciate the paradox that the driver presents. He tries to convince his audience that the profoundly complex object is neither irrational nor a thing to be abhorred.[4] Although gifted with a poet's fine sensibility, Henri has not transcended the dualist equation of rationality and harmony with order, irrationality and cacophony with chaos. He vastly prefers the laminar flow that leaves the speeding automobile unscathed, and fundamentally dreads the uncontrollable condition of turbulence. The driver's experiment challenges this dualist, and ultimately moralist, equation. What happens, he inquires, at that paradoxical moment of phase change from stable to chaotic? How do we express, moreover, the "utter harmony" that exists between these two states without prejudicially advancing the one over the other? Is the moment of impact, in which the perfect form disintegrates into formlessness, an ecstatic moment because the individual is relieved of the burden of self-preservation? The driver does not hold in contradiction his demand for "total destruction" and "this propensity of mine toward total coherence" (75). These absolutes ought to suggest a mutual exclusivity, and conventional wisdom would concur that they are not simultaneously obtainable. But that is in fact the paradox that the driver pursues, and the essence of chaos theory that Hawkes engages.

The driver—or Papa, as he refers to himself—broadens and embellishes the paradoxical nature of his proposition in a subsequent passage. He posits the possibility of encountering a farm truck on the dark road, a "miscalculation" that would result in "Disaster. Witless, idiotic disaster" (22–23). How nihilistic that their deaths should simply be the result of recklessness, a chance encounter that yielded no calculable results—a botched experiment. The driver is not so singularly in pursuit of the stochastic, random collision of molecules or automobiles as he is concerned with the relationship between the deterministic and the accidental: "what I have in mind," he declares to his unwilling audience, "is an 'accident' so perfectly contrived that it will be unique, spectacular, instantaneous, a physical counterpart to that vision in which it was in fact conceived. A clear 'accident,' so to speak, in which invention quite defies interpretation" (23). As a relatively superficial consideration of plot, Papa's desire to disguise the true nature of the "accident" is meant to deceive both his wife, Honorine, and quite possibly the insurance company, that other peruser of stochastic information. He means the simultaneous loss of husband, lover, and daughter to be far more devastating to Honorine (his route takes them past the chateau where she lies sleeping) than the revelation of a revenge plot could ever be—since the irruption of the uncontrollable in our lives is far more appalling than the vindictiveness of human tragedy.

Although the familiar silhouette of Alfred Hitchcock can be detected in the driver's proposition of something like the perfect crime, his theory more specifically addresses a conjunction of literature and science. The paradox of a perfectly contrived accident challenges the binary opposition between the deterministic and the stochastic, an event that has a causal relation to its antecedent and one that is purely the result of random processes. The driver has "conceived" the impact as testament to his own artistic "vision." The event will be a direct manifestation of his intentions, and yet be utterly disguised as the product of chance occurrence. Papa implies that the fine calculations that bring him to the stone wall—time, fuel, route—only serve to enhance the "unique, spectacular, instantaneous" impact and scattering of debris. The utterly unpredictable condition in which the destroyed automobile will be found—keeping in mind that the scene exists only as a projection and not a predictable surety—serves as an ineluctable "counterpart" of the design that brought it to pass. In this sense the driver joins those postmodern artists, composers, and writers who employ chance operations to sublimate authorial intention.[5] The device, the materials, and the mechanism are indicative of the artist's personal choice, but the resulting artwork is beyond the control of the artist. His "invention quite defies interpretation," intentionality becomes indistinguishable from happenstance, the cleverness of the criminal evades detection.

Hawkes describes the artistic challenge to himself and the driver—to conceive of the inconceivable—not in terms that are limited to such familiar media of creation as the sketchpad, notebook, or musical score but in terms that are more broadly applicable to the physical universe: "Death—cessation, annihilation—is the only thing I can think of that cannot be imagined. The only way that the artist-driver of the car can imagine it is through paradox. He conceives of the wreckage before it occurs; he recognizes that in destruction there is always a design for those of us who want to seek it; and he sees that in any design, any created thing, there is always the potential for the loss of its beautiful shape and its collapse into chaos" (LeClair, "Novelists" 28). The paradox of the contrived accident leads to a grander and perhaps universal paradox, the interpenetration of order and chaos. Hawkes claims as the result of his 128-page experiment the recognition that in a chaotic system it is possible to discover hidden order, in a scattering of debris it is possible to discover a pattern; and conversely, an orderly system or made object is susceptible to entropic—or catastrophic—decay, as the perfectly formed curve becomes distressed or destroyed. Hawkes demonstrates that the postmodern novelist apprehends one of the central tenets of chaos theory not as a figure of speech but as a physical principle.

Because the driver's preconception of the wreckage, the aftermath of his own annihilation, is at issue in Hawkes's argument, it would be best to visit the site of this "private apocalypse" as it is described—hypothetically—in the narration:

> we have at the outset the shattering that occurs in utter darkness, then the first sunrise in which the chaos, the physical disarray, has not yet settled—bits of metal expanding, contracting, tufts of upholstery exposed to the air, an unsocketed dial impossibly squeaking in a clump of thorns—though this same baffling tangle of springs, jagged edges of steel, curves of aluminum, has already received its first coating of white

> frost. In the course of the first day the gasoline evaporates, the engine oil begins to fade into the earth, the broken lens of a far-flung headlight reflects the progress of the sun from a furrow in what was once a field of corn. The birds do not sing, clouds pass, the wreckage is warmed, the human remains are integral with the remains of rubber, glass, steel. A stone has lodged in the engine block, the process of rusting has begun. And then darkness, a cold wind, a shred of clothing fluttering where it is snagged on one of the doors which, quite unscathed, lies flat in the grass. And then daylight, changing temperature, a night of cold rain, the short-lived presence of a scavenging rodent. And despite all this chemistry of time, nothing has disturbed the essential integrity of our tableau of chaos, the point being that if design inevitably surrenders to debris, debris inevitably reveals its innate design. (58–59)

One appreciates the fact that this passage moves inductively from a multitude of fine details to a statement of principle; the nature of chaotic behavior makes it irreducible to a normalized statement. Because this "tableau of chaos" is complex, it is rich in information that must be recorded before the system itself can be understood.[6] The driver is determined—he plans—to shatter his elegant sports car in the field of an abandoned farm: the furrows of the field, the marks of human cultivation and linear order are themselves undergoing erasure with the "chemistry of time"; so the disarray of mechanical parts becomes one with the landscape, the background, from which order arises and into which order decays. One must abandon the anthropocentric prejudice that finds loathsome that which threatens to absorb the products of human manufacture. In this chaotic scene, the driver does not see gruesome fragmentation but an "essential integrity" amidst the dispersal of car parts. Further, the natural and mechanic elements have become thoroughly interspersed: with "a stone lodged in the engine block, the process of rusting" begins, so that the man-made metallic object, which had been artificially sealed in an oil-filled casement, now interacts, mixes, with other natural substances. "The broken lens of a far-flung headlight" that once artificially concentrated a beam of light now "reflects the progress of the sun," and so participates in a recursive order on an infinitely more complex level. The driver complains, through his description of the wreckage, that we have been hermetically sealed within a compartment of petty orderliness in which any fissure is appalling, at the expense of a far greater participation in the actualities of the physical world. Thus the collapse of that petty order which is the sports car so lovingly maintained, reveals an order on a different scale, or of another sort.

Connoisseurs of Chaos

In literature, science, and philosophy a paradigm shift occurs, late in the twentieth century, in the conception of the relation between order and disorder. In the modernist period—although certainly this impulse is identifiable at other prior moments in literary history—the function of the artist was to impose an order, self-consciously, through the use of artifice and design, upon an otherwise disordered, fragmented, and recalcitrantly irrational world. Modernist artists were aware that the patterns they imposed were impermanent assertions of their creativity against the dissolution that surrounded them. In their conception, chaos presents itself as the antithesis of order. Wallace Stevens, for example, addresses this problem from the modernist perspective in some of his most important work, and in no simple manner. In "Imagination as Value," he states: "imagination is the power that enables us to perceive the normal in the abnormal, the opposite of chaos in chaos" (*Necessary Angel* 153). Stevens's assertion seems valid: that reason alone does *not* permit us to perceive the order in chaos, and that the attorneys and insurance executives whose profession he shared considered the imagination to be abnormal, if not irrational. In his tribute to the powers of the imagination, Stevens is much like Hawkes in that the artistic imagination permits him to perceive the "impossible object." But unlike Hawkes, Stevens places an exceptional emphasis on the mind's ability to establish order, in opposition to chaos. Thus, in "Anecdote of the Jar," a kind of aesthetic parable, the jar as a figure for the ordering capacity of the mind lends shape to the "slovenly wilderness" (*Collected Poems* 76). And in "The Idea of Order at Key West," he acknowledges and approves the "maker's rage to order" the turbulent chaos of the sea (*Collected Poems* 130). For Stevens, the human mind creates meaning in an otherwise irrational and indifferent physical universe.[7] His position is entirely in accordance with Roland Barthes's description of the modern consciousness as an imagination "fabricating meanings" in the world (*Critical Essays* 218). The modern artist is a maker, a fabricator of meanings and orders that must, as Stevens says elsewhere, "suffice." The critical attribute of the modern artist remains the power to impose order on an inchoate world.

Returning to the more expansive genre of fiction, one sees that the paradigm in place in the modernist novel is much the same as it is for Stevens. In their overview of the situation, John Fletcher and Malcolm Bradbury argue that the modernist, or introverted, novel aspires "to the condition of ordered game," cognizant as it is of "the fact that all human order is an imposition on the absurdity of experience." Granted that human order is both arbitrary and impermanent, they conclude: "Hence *the novel is implicitly design and design only*, a form of art and joy, a world pleasured by its own making. Reality itself being offensively literal or hideously surreal, the novel has thus tended to become, for a substantial and international group of writers, a particular occasion of order—or disorder—to set against all other orders, or disorders" (412–13, my emphasis). Fletcher and Bradbury are certainly correct in their description of the goals and methods of the modernist novel. But as they intend this description to be applicable as well to postwar fiction, some emendation is in order. These two critics iterate the modernist emphasis on "making," and the "imposition" of human order on an abhorrently disorganized world. The modernist novel may well be "design and design only," but Hawkes demonstrates that the postmodern novel must relate "design and debris" as a fully integrated system. If we can return to the driver's "theory," he states that "ours is the power to invent the very world we are quitting" (57). The postmodern artist proposes a world pleasured both by its making and its unmaking, its creation and its destruction, its invention and its abandonment.[8]

Hawkes transcends the dualist imposition of order on chaos that is characteristic of a fabricatory modernism, in which the artist must inevitably be frustrated by the impermanence of his creation. Those critics who insist on a dualist interpretation of the novel thrust it back into the paradigm of modernism. For example, Eric Henderson claims, "It is Hawkes himself, through linguistic and stylistic devices, through the structuring capacity, who exhibits the authorial (and authoritarian) need to impose a sense of order and coherence on a world characterized by random eruptions of violence and chaotic disturbances" (3034A).[9] Rather, Hawkes wishes to sustain the paradox of design *and* debris and to assert—as the driver so clearly states—the principle of the interpenetration of order and chaos: again, "the point being that if design inevitably surrenders to debris, debris inevitably reveals its innate design" (59). As a postmodern novelist, Hawkes does not shrink before the proposition of "unmaking" or decreative force; he extols the complementarity of the two terms; and finally, he proposes the existence of an orderly disorder. In his statement of the complementarity of design and debris, Hawkes might well be echoed by Douglas Hofstadter who observes, "It turns out that an eerie type of chaos can lurk just behind a façade of order—and yet, deep inside the chaos lurks an even eerier type of order" (Gleick, back cover). Hawkes's *Travesty* illustrates the tenuousness of authoritarian control as it slips into madness, the fragility of pattern as it dissolves into irregularity; and it proposes the revelation of some hidden order in the scatter of random occurrences, some more profound design within the welter of chaos.

Recursive Symmetry in Narrative

Michel Serres, whose career has been devoted to the cross-disciplinary study of science and the humanities, provides confirmation of the paradigm shift as it is described in Hawkes's novel. Serres deplores the "dualist hell" that has separated science and literature, reason and imagination, order and disorder, unity and multiplicity. The result of such dualism is a "desire to dominate. To think in terms of couples of separate concepts is to prepare fearsome weapons —swallow-tailed arrows and darts—so as to control the space and to kill" ("Dream" 233). Instead he observes that everywhere in the physical universe unities are combined with scattered multiplicities: "the mixture is so intimate that there is not one chaotic or distributional area that is not surrounded with systems. . . . Nor is there one area of system that is not surrounded with distributions. . . . It is as if the essential concern here were to border disorder with order and unities with multiplicities. These neighborings are so refined that the interpenetration is total" ("Dream" 231–32). Like the most elaborate of parquet floors, the inlay of aspects of order and disorder in the physical universe is so complete that the contrasts between the separate materials are subsumed within the interaction of the whole. Serres professes, "I believe that they have been sown into a sporadic jigsaw puzzle. This reciprocal plunge, this bathing of islands of order or negentropy in a fractal sea of rumbling, this bathing of lakes of noise in a formerly glacial earth, is not endowed with regularity, for it is itself a distribution, a sheer multiplicity." However sporadic the reciprocity, "the universe vibrates endlessly between the two," pulsating between order and disorder ("Dream" 232). This acknowledgment of the interpenetration of order and chaos is the principal revelation of the driver's monologue in *Travesty* and an essential component of Hawkes's postmodernism.

The narrator of *Travesty* is quite articulate regarding these themes. Very near the beginning of the novel, Papa "replies" to Henri, "Slow down you say? But the course of events cannot be regulated by some sort of perversely wired traffic policeman." He explicitly rejects a Newtonian determinism that could predict the result of an action with mechanical precision. The variables are too many for such sureties: "Our speed is a maximum in a bed of maximums which happen to include: my driving skill, this empty road, the time of night, the capacity of the car's engine, the immensity of the four seasons lying beyond us between the trees or in the flat fields" (15). Each of these conditions is unpredictable, all have been pushed to their very limits, yet they are so finely interrelated that the failure of any one is the failure of the system. Papa concludes, "Like schoolboys who have studied the solar system (I do not mean to be condescending or simpleminded) you and I know that all the elements of life coerce each other, force each other instant by instant into that perfect formation which is lofty and the only one possible" (15). Such perfection is not achievable through the regulatory impulse of a single determining actor, but rather through the infinite adjustment and interaction of all elements involved.

The driver continually interrelates choice and chance, plan and accident, artificial and natural, order and chaos, until his audience is forced to surrender the fond notion that any decision, act, or result in their life is purely a matter of either intentionality or random occurrence; these things intertwine in the course of our passage. The driver is missing a lung, attributable to "the war of course." Is such a condition accidental or intentional? "At any rate," he continues, "it is probably true that my missing lung determined long ago my choice of a doctor." The doctor, in turn, is missing his left leg, "amputated only weeks before the poor fellow's wife ran off, finally, with her lover of about twenty years' standing" (26–27). The driver deliberately links his choice of a physician—when others shun the man—to the concatenation of disorder in his life. Are these misfortunes a matter of coincidence, or is the final wreckage of the doctor's marriage instigated by his physical disability? In either case, Papa is attracted to debris:

> The affinity is obvious, obvious. But by now you will have perceived the design that underlies all my rambling and which, like a giant snow crystal, permeates all the tissues of existence. But the crystal melts, the tissues dissolve, a doctor's leg is neatly amputated by a team of doctors. Design and debris, as I have said already. Design and debris. I thrive on it. For me the artificial limb is more real, if you will allow the word, than the other and natural limb still inhabited by sensation. (27)

The driver's design, his planning and deliberation, leads him to destruction and debris; that disorder, in turn, leads him to a form of organization that is perhaps "more real" than that which he had previously perceived.

The driver's choice of the "snow crystal" illustrates the way in which order arises out of seeming chaos, a plan emerges from ramblings, and then as quickly melts into disorder. The snow crystal provides an apt symbol of order in chaos because it is a familiar example of fractal structure. Gleick points out that "the word *fractal* came to stand for a way of describing, calculating, and thinking about shapes that are irregular and fragmented, jagged and broken-up—shapes from the crystalline curves of snowflakes to the discontinuous dust of galaxies. A fractal curve implies an organizing structure that lies hidden among the hideous complication of such shapes" (113–14). Hawkes's narrator has clearly invested his confidence in the principle of regular irregularity, in which "the degree of irregularity remains constant over different scales" (Gleick 98). If patterns reveal themselves in the jagged shapes of snowflakes and galaxies, surely an underlying design will be revealed in the hideously fragmented remains of the automobile: what pertains to the microscopic and the macroscopic should also pertain to the middle distance of human affairs. The "giant snow crystal" is symbolic of a fractal ordering at every scale, permeating every level of existence.[10]

Perhaps the defining quality of fractal order is self-similarity. Gleick explains, "Self-similarity is symmetry across scale. It implies recursion, pattern inside pattern." Fractal shapes not only "produce detail at finer and finer scales," they also produce "detail with certain constant measurements. Monstrous shapes like the Koch curve display self-similarity because they look exactly the same even under high magnification" (103). Postmodern fiction is no stranger to the principles of recursive structure or infinite regress as illustrated by the Koch curve. The world-within-a-world, or the Chinese box model, has served Borges, Burroughs, and Barth among others.[11] No other worlds are posited by the narrator in *Travesty*—the one he enacts is complicated enough—but at least one critic, Patrick O'Donnell, has remarked on the intensive degree of "symmetries, repetitions, identifications, and recurrences" in so short a novel (139). The "affinity" between the single-lunged driver and his one-legged physician is one such detail that prompts the existential query of multiply receding reflections, as when one stands between two mirrors. At a "finer scale" in the narrative, Papa recounts a story told by Lulu, a resort manager, to his wife Honorine about a dwarfish man, "the possessor of a left arm nipped off and drawn to a point at the elbow by one of those familiar accidents of birth," sent by his demanding mistress to sell a branch of mimosa in the street. Here we encounter a "nesting" of stories (Papa's direct discourse, within which the *Chez Lulu* episode, within which Lulu's tale) in which the "constant measurement" of a malformed body appears at a further reduction of scale. It is surely no accident that the pitiable man is humiliated when "a small but elegant automobile drove past with an enormous heap of gleaming, yellow mimosa covering its entire roof" (112–13). Such similarities across scale may be taken either as an indication of fictional artifice or as the revelation of pattern underlying chaos. In any case, the affinity of the inset story with the driver's own inability to please his wife, and his mocking, destructive solution to the problem, is indeed obvious.

Further mirrorings occur within the novel, notably the attribution of

Henri's creative gift as well as his psychosis to Papa. Such similarities between characters are provocative, especially when encountered in profusion.[12] But the most important of these mirrorings is the driver's allusion to another, possibly fatal accident that he describes as "a singular episode of my early manhood. . . . Certainly it determined or revealed the nature of the life I would lead henceforth as well as the nature of the man I had just become. It is something of a travesty, involving a car, an old poet, and a little girl" (47). For Papa, the earlier incident determines the course of his life; it is a random occurrence that coalesces into—or reveals—a governing pattern. The episode is his youthful "moment of creativity" (47) that underscores the particularly destructive nature of the driver's artistry. Hawkes intends the account of this "travesty" to function as a self-similar model for the novel as a whole, replicating its primary characters and scenario. Only the nodding, inattentive reader can fail to appreciate the "travesty" within *Travesty*. Not until the very close of the novel does the driver describe this "formative event" (125) of his life in detail. In a similarly powerful sports car, the driver believes that he may have struck a beautiful, dark-haired girl (who, in her apparent ingenuousness, mirrors Chantal) being led down a crowded street by an old man with the air of a poet (thus resembling Henri). Irritated by the possessiveness of the old man, the driver deliberately accelerates, brushing past the girl. Because he refuses to look back, Papa remains unsure as to whether he strikes her or not. The uncertainty of this scene, as much as the replication of characters, indicates its similarity to present circumstance, as the three are suspended in the process of a "contrived accident" without clear and definite result. If this "formative event" in the driver's life serves as a prototype for his current actions, the entire novel is a replication on another scale. In metafiction, these incidences of recursiveness usually provoke ontological questions: how does one distinguish between the fiction and the reality? But in *Travesty* the many instances of self-similarity and recursion are there to confirm for the driver his belief in an underlying, permeating design.

It is also likely that one tragedy in the chronicles of postwar French literature serves as the "formative event" for Hawkes's *Travesty*. Of the two epigraphs to the novel, one is by the novelist and playwright in the Theater of the Absurd, Albert Camus. The excerpt from his novel, *La Chute* (1956; *The Fall*) reads, "You see, a person I knew used to divide human beings into three categories: those who prefer having nothing to hide rather than being obliged to lie, those who prefer lying to having nothing to hide, and finally those who like both lying and the hidden. I'll let you choose the pigeonhole that suits me." Of the personnel in Hawkes's novel, one can speculate as to which category they fall into; I am inclined to place the driver in the last, for his penchant for conflating truth and falsity, intention and accident. Camus's biographer, Olivier Todd, provides the details of his unexpected death in an automobile accident in 1960 in the company of the Gallimard family, Janine, Michel and Anne:

> They left the next morning, drove for three hundred kilometers, and had a light meal at Sens. The Gallimards were teasing Albert about all his woman friends, and he replied that he managed to make them all happy. Michel was at the wheel, and Albert was beside him. Janine had given up her place in the car saying, "You're taller than I am."
>
> Twenty-four kilometers outside Sens on the Nationale 5 road, between Champigny-sur-Yvonne and Villeneuve-la-Guyard, the Facel-Vega swerved, went straight off the road, slammed into a plane tree, bounced off another tree, and broke into pieces. Michel was seriously injured, Janine and Anne were unharmed, the dog disappeared, and Albert Camus was killed instantly.
>
> The dashboard clock, which had been thrown into a nearby field, was stuck at 1:55 p.m. Camus had often told friends that nothing was more scandalous than the death of a child, and nothing more absurd than to die in a car accident. (413)

That a writer so committed to the premise of the inescapable finality of death, and so convinced of the absurdity of an accidental death, should die in this manner is the unspoken travesty of Hawkes's novel.[13] The unexplained swerve, the dashboard clock lying in the field, and the young girl (who is eighteen) exist in yet one more mirrored dimension between fact and fiction.

Mastery and Submission

The principle of self-similarity encourages us to consider the interrelation of characters in *Travesty*, but it is at least as valuable to consider the distinctions that separate them. Within the confines of the automobile, Hawkes presents an equation in three variables, a triangulation of force and motive. N. Katherine Hayles advances the argument that "chaos has been negatively valued in the Western tradition" because of the "predominance of binary logic in the West. If order is good, chaos is bad because it is conceptualized as the opposite of order." Essentially in agreement with Serres's invective against the "dualist hell" of binary opposition, Hayles proposes that "not-order is also a possibility, distinct from and valued differently than anti-order" (*Chaos and Order* 3). Let's consider, then, the application of these three values, order, anti-order, and not-order with respect to Hawkes's distribution of character traits. Henri urges the driver to return to his senses, to be "reasonable." He reasserts the most conventional sort of orderliness in the Western tradition. There is, understandably, a tinge of desperation in his pleas, a patent recognition that the assertion of reasonable behavior is never more than an agreed-upon standard; the façade of civilized order crumbles away. The driver reveals this disintegration when he accuses Henri of being "only the most banal and predictable of poets. No libertine, no man of vision and hence suffering, but a banal moralist" (14). The driver's disregard for order as safe-haven is easily surpassed by his contempt for the pretensions, the presumptions that the orderly man displays. Henri's career as a poet is founded upon his having overcome, mastered, the psychological distress that once afflicted him. People admire Henri, like T. S. Eliot and other modernist poets, for his "desperate courage," and especially for "having discovered some kind of *mythos* of cruel detachment" (43). Hawkes seeks to embarrass those who would lionize such a clearly masculinist control and its mystique of impersonality. Henri epitomizes an order that is unemotional, indifferent, unchanging; it aspires to permanence and totalization. But the driver demonstrates how subject to cracking, blistering, and fading such a wash of permanence can be. Thus the modernist emblem of order is held up as a falsehood.

Chantal's immediate reaction to her father's announcement of his intentions is to flail the driver about the head and shoulders and then collapse behind his seat sobbing uncontrollably. She is the figure of anti-order, of entropic decay in the novel. She falls apart. Chantal has also been dubbed the "porno brat" by her mother because "she was forever stumbling into the erotic lives of her parents" (12–13). And of course, her precocious sexuality provides for further disruption of Papa's erotic life when she takes Henri for a lover. The lesson implied by this, that sexuality is destructive, is altogether too simple. The driver describes the effect of his actions on the two women in his life: "But Chantal and Honorine—what a pair of names. And to think that at this instant the one is white-faced, tear-streaked and clinging to the edge of hysteria in lieu of prayer directly behind us, while the other sleeps in the very chateau we are approaching. But be brave Chantal. There will be no comforting Honorine when she receives the news" (13). What a pair of names indeed! The driver is self-consciously aware—one might say metafictively aware—that these women represent the relegation of uncontrollable emotion and destructive sexuality to the feminine. In the binary logic of the Western tradition, chaos as a purely destructive force and the hysterical woman are members of the same paradigm. In classical terms, the feminine is figured as erotic enchantress (Chantal, Circe) or as goddess-like object of devotion with the power to destroy her suitors (Honorine, Helen). They are thus conventions of a feminine chaos that must be contained or controlled for the persistence of an orderly society. But just as the driver is aware of the falsity of Henri's prestige of detachment, so he also recognizes that the association of chaos with hysteria is an old prejudice. Papa wishes to redefine the nature of chaos as something much more complicated and profound than anti-order. Chantal and Honorine are indeed unbelievable.

The driver, then, is the figure of not-order, racing toward that point at which design and debris become interchangeable. To do that he must first challenge those binary oppositions that are so dearly held within the culture. His principal method in this endeavor is a process of exchange. Binary opposites should be clearly defined in their differences. But what if the properties of one character are usurped by another? Papa exchanges the bruised ego and jealousy of the cuckold for the imaginative creativity of the artistic mind: "we have one more scrap to toss on the heap of our triumphant irony. Because in our case it now appears that the poet is the thick-skinned and simple-minded beast of the ego, while contrary to popular opinion, it is your ordinary privileged man who turns out to reveal in the subtlest of ways all those faint sinister qualities of the artistic mind" (100). Papa thus attributes to the principle of not-order the ability to create as well as to destroy; like the artist, he assumes the capacity to bring something out of nothing. Ilya Prigogine and Isabelle Stengers describe this aspect of the paradigm shift in chaos theory, pointing out that "non-classical science has taught us that trajectories can become unstable and that stochastic chaos can become creative. In certain circumstances, evolution bifurcates, the homogeneous disorder is no longer stable, and a new order of organized functioning is established." They distinguish between an entropic "indifferent disorder" and the "strange tumult" that arises in instances of "creative chaos" ("Postface" 153). The driver thus describes this paradigm shift that attributes an artist's creativity to "not-order."

There is also a degree to which Papa emulates the sexual abandon of his daughter. Papa acquires his mistress Monique when she was exactly the same age as Chantal (about twenty) when Henri "determined to extend to her the love of the poet." Despite his denial that "I was trying to duplicate my daughter in my mistress," Papa is obviously pleased that Monique resembles Chantal in her diminutive size (65). The salient detail in the account of this rather risqué episode is the discovery by Papa of a mild degree of sadomasochistic passion. He feels "suddenly inspired" (the terms are once again artistic) to spank Monique, eventually becoming aroused (68). Monique, however, is less than thrilled by her paddling and retaliates with her garter belt, delivering "full in the lap the pain of the little metal grips" (73). The episode ends with this lesson for father (and by extension, for the artist): "I learned that I too had a sadistic capacity," and that "it reveals that I too have suffered and that I am not always in total mastery of the life I create, as I have been accused of being" (74). Papa learns of the dissipation of sexual obsession, the loss of control that accompanies it. The "total mastery" that he in fact is currently wielding over the passengers in the car can erode, be converted into submission or slavery. And in fact that control does dissipate at the moment the car crashes. The driver, as the figure of "not-order," articulates both the dissipative energies of the sexual deviant Chantal and the creative organization of the poet Henri. The binary opposition between order and anti-order becomes blurred, is itself unmade, as it is interrelated in the character of the driver.

The Swerve

Like the climactic act of violence that in the Greek theatre always occurred offstage, the critical action in the narrative of *Travesty*—the swerve that sends the sports car toward collision—is anticipated and its repercussions analyzed, but it remains beyond the immediate experience of the narrator and his audience. The driver has carefully chosen the site of the collision. The swerve that propels the car toward impact is deliberate and thus distinguished from the random accident that would consign the occupants of the car to "the toneless world of highway tragedy" (11). The driver rejects the old Roman viaduct over which they must pass, "that narrow dead viaduct that spans the dry gorge and always reminds me of flaking bone" (23), as an appropriate site. Because the viaduct is a *memento mori*, it is entirely too obvious: "All those 'logical' details and all those lofty 'symbols' of melodrama speak much too clearly to the professional investigator" (24). No, although the swerve is determined by the driver, it must in every way appear to be an inexplicable, unmotivated occurrence. Instead, beyond the viaduct, the car "shall make an impossible turn" toward the abandoned farmhouse. The driver proposes a "turn that is nothing less than incomprehensible," leading to a collision that is both "severe and improbable" (24–25). The swerve, the deviation from the road, remains incomprehensible to the authorities precisely because it defies the logical, symbolic order intrinsic to human affairs. It is madness, it is brilliance. It "must be senseless to everyone except possibly the occupants of the demolished car" (25). The "incomprehensible" swerve and the "improbable" crash confound those investigators—or for that matter, those critics—who wish to interpret the impact either as an error in driving skills or as a moral failing in the act of murder-suicide. The authorities hear only a noise which obscures the meaning of the incident, stymies their attempt to distinguish between accident and intention, and so they feel compelled to squelch that interference in the signal. But such a noise is paradoxically the instigation of a rather profound communication among the "victims." Only a slight extension of the driver's statement allows us to interpret the crash as a noisy channel for a message between Papa as sender and his presently widowed wife Honorine as receiver.

In "Lucretius: Science and Religion," Michel Serres gives particular attention to the concept of the *clinamen atomorum*, or swerve of the atoms, which Lucretius introduces in his verse treatise on physics, *De Rerum Natura*, to explain the origin of the cosmos from an earlier, undifferentiated state of atomic flux. The atoms "fall" through space in a parallel and undisturbed motion until, as Lucretius suggests, "at utterly unfixed times and places they swerve a little from their course, just enough so that you can say that the direction is altered. If the atoms did not have this swerve, they would all fall straight down through the deep void like drops of rain, and no collisions would occur nor would the atoms sustain any blows. Thus Nature would never have created anything" (*On Nature* 47). For Lucretius (elaborating on Epicurean philosophy) the stochastic swerve in matter ultimately enables free will in human beings, "by virtue of which we each go where pleasure leads us and, like the atoms, swerve in our courses at no fixed time and in no predetermined direction, but when and where the mind itself impels" (*On Nature* 48). More important to Serres is the fact that Lucretius attributes the self-organizational power of the cosmos, the creative force of Nature, directly to the random swervings of the atoms and to the turbulence that their collisions provoke. For those of his readers who still required a godly image, Lucretius attributes the pleasure motive and the factive ability of turbulence to Venus. Serres defines the *clinamen* as the "minimal angle to laminar flow [that] initiates a turbulence" (*Hermes* 99). He associates the initial condition of laminar flow with the deterministic physics "of repetition, and of rigorous trains of events. . . . There is nothing to be learned, to be discovered, to be invented in this repetitive world, which falls in the parallel lines of identity. . . . It is information-free, complete redundance" (*Hermes* 99–100). The physics of classical dynamics—here associated with Mars—describes a world of rigid causality: "Without the declination, there are only the laws of fate, that is to say, the chains of order" (*Hermes* 99). Such a martial physics will inevitably result in "death at the end of entropy" (*Hermes* 100).

The introduction of the swerve, however, occurs (in the words of Lucretius) *incerto tempore . . . incertisque locis* (Leonard and Smith 333; II 218–19), at uncertain times and places; it is thus indeterminate and unpredictable, an aspect of random dispersion. Serres argues that "the angle interrupts the stoic chain, breaks the *foedera fati*, the endless series of causes and reasons. It disturbs, in fact, the laws of nature. And from it, the arrival of life, of everything that breathes; and the leaping of horses" (*Hermes* 99). Serres may seem to make a dramatic leap from atomic motion to bio-evolution, but the collision of atoms and the resulting turbulence allow for increasingly complicated structure that an unchanging, parallel motion of atoms does not permit. The swirling commotion of turbulence actually creates pockets of negentropy, small areas of increasing organization, within the large-scale flux. Prigogine and Stengers, reading Serres's reclamation of Lucretius for contemporary physics, point out that "while turbulent motion appears as irregular or chaotic on the macroscopic scale, it is, on the contrary, highly organized on the microscopic scale. . . . Viewed in this way, the transition from laminar flow to turbulence is a process of self-organization" (*Order Out of Chaos* 141–42). Referring to the Lucretian model, they reinforce Serres's position:

> the infinite fall provides a *model* on which to base our conception of the natural genesis of the disturbance that causes things to be born. If the vertical fall were not disturbed "without reason" by the clinamen, which leads to encounters and associations between uniformly falling atoms, no nature could be created; all that would be reproduced would be the repetitive connection between equivalent causes and effects governed by the laws of fate (*foedera fati*). (*Order Out of Chaos* 303)

The swerve that marks the transition from laminar to turbulent flow ushers in a generative rather than a static world, a world that is information-rich rather than information-free, a world that moves toward a greater degree of complexity rather than the death that the laws of entropy ultimately demand.

One might think that *Travesty* deviates from the Lucretian model because the swerve is both premeditated and in a predetermined direction. Much of the novel's discourse is devoted to establishing the motive and causation of the swerve, as an expression of the driver's deliberate intention rather than as an impulsive gesture. The swerve thus constitutes an element of plot; it is not merely an aspect of story. Because the driver apparently intends the automobile's swerve to inflict death upon all occupants, one wonders how closely Papa's philosophy could be allied to the Lucretian physics of Venus instead of Mars. Or is the deviation so very great? The driver's argument throughout the novel has been to redefine the nature and relation of order and chaos, design and debris. He has sought to dispel the commonly held belief in chaos as a purely destructive force and order as a preeminently sustaining structure. With Lucretius and Serres, the driver conceives of laminar flow, straight-line travel, as resulting inevitably in "death at the end of entropy." For Papa—the admirer of fast cars—there is greater misery in the slow, sure decay of the condition in which he finds himself. But the turbulence brought on by the swerve and collision promises the possibility of an increase in complexity. The social order of family and friend has clearly failed; perhaps there is an innate design, a more complex array, that will reveal itself in farmhouse field. There are new worlds, other possible combinations of matter. Ultimately, the aspect of indeterminacy in the swerve of *Travesty* lies with the reader's inability to ever know for sure whether the promised turn does come.

Hawkes is intent on confounding the "professional investigator" who sees in the swerve and the resulting turbulence only death and disintegration. In *Travesty*, the death of the narrator—and hence the death of the novel—signals the creation of a more recondite order. Logical details are scattered across a furrowed field, cause and effect is obscured in the absence of skid marks, and the chains of a grim self-preservation are broken on impact. Like Serres, Hawkes believes that it is not the meager reassurances of a rationally derived order but the revelation of innate design in turbulence that ultimately sustains life and being.

3

Discipline and Anarchy

Disrupted Codes in Kathy Acker's *Empire of the Senseless*

I like a look of Agony,
Because I know it's true—
—Emily Dickinson, Poem #241

Discipline organizes an analytical space.
—Michel Foucault, *Discipline and Punish*

Good authors, too, who once knew better words,
Now only use four-letter words
Writing prose.
Anything Goes.
—Cole Porter, *Anything Goes*

A double-edged dagger, hilt up, on which is impaled a whorled and leafy rose. The rose, a heart, bleeds droplets. Around the blade and point are intertwined banners that read "Discipline and Anarchy." This hand-drawn figure appears at the close of Kathy Acker's *Empire of the Senseless* (1988), a book which she dedicates to her tattooist. Although this tattoo design is inscribed on the back of the book, the pain of its needle and the poison of its ink are felt throughout the novel. Acker drives the sword of pain through the rose of pleasure; she binds the cords of discipline and unties the knot that restrains anarchy. The iconography suggests that the competing principles of discipline and anarchy, intentionality and impulse, control and freedom are inextricably linked in the novel (Figure 2). The Pandora of postmodernism releases a flurry of evil and disease into a culture ready to receive it. Acker's writing is anarchistic to the core: she plunders the cultural storehouse of Western literature, liberating the classics through plagiarism; she violates every known taboo, revels in obscenity, smashes genre rules, and commits violence on her

Figure 2. Acker's Tattoo

characters that would make the Marquis de Sade blanch. But like Sade, she displays a penchant for discipline as control as well as punishment. Anarchism runs its course without resistance, entropically feeding on the fuel of stale and repressive social order until that fuel is exhausted. And discipline carried to any restrictive extreme will at last inspire revolt.

With the pain of her own writing in mind, Emily Dickinson declared that "The Attar from the Rose . . . is the gift of Screws" (Poem #675; *Final Harvest* 171). The belle of Amherst was no stranger to agony and violence in her work; literary form can only be painfully "expressed." The writing of *Empire of the Senseless* more vigorously courts anarchy. The reader may well be appalled by the mayhem encountered on any page. But the text finally deports itself in a disciplined fashion. Acker claims in an interview with Ellen G. Friedman that "it was the structure [of *Empire*] that really interested me—the three-part structure" ("Conversation" 17).[1] The question nags as to why a text so engrossed with disorderly behavior would be preeminently interesting to its author for its structure. Acker the plagiarist might want to lift a line from that most patriarchal poet of suburban Connecticut, Wallace Stevens: "This is form gulping after formlessness" (*Collected Poems* 411). She envisions her novel as a triptych with a rather deliberate progression of effects: the deconstruction of the patriarchal order in "Elegy for the World of the Fathers," the liberation that follows from an end to repression and inhibition in "Alone," and the formation of a new society on the very ground of transgression in "Pirate Night." Acker provides the structure's rationale as follows:

> The first part is an elegy for the world of patriarchy. I wanted to take the patriarchy and kill the father on every level. And I did that partially by finding out what was taboo and rendering it in words. The second part of the book concerns what society would look like if it weren't defined by oedipal considerations and the taboos were no longer taboo. I went through every taboo, or tried to, to see what society would be like without these taboos. . . . The last section, "Pirate Night," is about wanting to get to a society that is taboo, but realizing it's impossible. ("Conversation" 17)

I want to discuss these three effects and the turbulence that accompanies each in turn. But Acker's statement reveals something important about her methodology as an artist.[2] As the princess of violation and disruption, Acker can be expected to disregard the traditional rules of fiction. But an aimless thrashing of the novel form does not follow. Instead she claims adherence to the method of conceptual artists such as David Antin and William Burroughs. In an interview with Sylvère Lotringer, Acker argues that "Form is determined not by arbitrary rules, but by intention." Her emphasis has been "on conceptualism, on intentionality" ("Devoured" 3). On the one hand Acker rejects the plotting of conventional fiction writing and the proceduralism of some other avant-garde novelists as overly constrictive. On the other hand she finds reliance on impulse and intuition alone to be inadequate to her task. She writes "by process" (4), but finds the form of the novel in conceptualism. In this way the event of the novel as a whole can be orderly even though its many actions may be chaotic. Acker's methodology thus weds impulse and intention. She envisions the collapse of the patriarchal order into a state of liberating and enabling anarchy as a function of the book's structure. Her methodology is directly supportive of the concept of *Empire of the Senseless*, which is not unlimited anarchy but the intrinsic relation between discipline and anarchy, as the banner on the tattoo declares.

The Decline of Order: Sadomasochism

In "Elegy for the World of the Fathers" Acker contemplates the utter corruption of the patriarchal order, attacking monotheism, capitalism, the phallic power bestowed by Freud in the Oedipal myth, marriage as a "collective crime" (*Empire* 7), and the relegation of the sexual body to a commodity.

Acker's female protagonist, Abhor, is described as "part robot, and part black" (3). The character's name and her cybernetic-minority status (derived from William Gibson's *Neuromancer* [1984])[3] suggest the contempt, the abhorrence that patriarchal society has traditionally bestowed upon women, half-breeds, and the disenfranchised poor. Acker deliberately neglects to develop her character's status as robot or black, which suggests that these are political identities—"sub-human" designations—against which Abhor must struggle as an individual. The initial chapters purport to provide the protagonist's genealogy and early years, a gesture toward the *bildungsroman* which Acker quickly deconstructs. Part I of the novel is preceded by a tattoo drawing of a skull and rose branch bearing the legend "My Family Fortune." Abhor's paternal grandmother is a German-Jewish refugee fleeing Nazi persecution and the father-führer; upon her deliverance in Paris, she is immediately forced into child-prostitution. For Acker, who documents her performances in Times Square sex-shows during the 1970s, any sex within the patriarchal system becomes equated with money. Or, as literally stated by Abhor, "Sex was joined to money" (9).

Acker wastes no time exploring the primal taboos of incest in the first chapter, "Rape by the Father." Abhor is already a victim of patriarchal moralism: "I knew I was evil cause I was fucking" (11). Her teenage sexual activity inspires Daddy's transgression. Her appeal (by telephone) to her mother hardly delays father's actions because the female lacks the authority to enforce a taboo within the patriarchal order. Abhor's reaction to her violation is typically complex: "Part of me wanted him and part of me wanted to kill him" (12). Western culture has of course punished women for both responses—admitting desire for the father, and patricide. Acker explores the divergent and yet related responses of submission (desire) and revolt (hatred). In trying to enforce taboo the patriarchy has declared "Neither. Nor." Abhor's release from that order is her willingness to contemplate both the unthinkable desire for her father and the resentment for his unspeakable violation of her. She sees her oppression and her freedom in the same figure.

Daddy's transgression of the moral code may not have been salutary for Abhor, but it makes an important contribution to her concept of social order:

> Daddy left me no possibility of easiness. He forced me to live among nerves sharper than razor blades, to have no certainties. There was only roaming. My nerves hurt more and more. I despised those people, like my mother, who accepted easiness—morality, social rules. Daddy taught me to live in pain, to know there's nothing else. I trusted him for this complexity. (10)

The moral code enforces an absolute distinction between "right" and "wrong." It rewards conformity with "easiness," an illusory sense of self-satisfaction. The world, approached from comfortably within these "social rules," is simple. But for Abhor the simplicity of the moral code and those who adhere to it are despicable. Rather than endure the restrictions and submit to the governance of the patriarchy, she prefers to dwell in uncertainties and in pain. Abhor recognizes that to defy the dominant order is to invite discomfort, insecurity, and conflict. But no matter how much the dominant order rewards conformity, that in itself does not give it sole purchase of the truth in reality. The dominant order always seeks to suppress conflicting systems, ambiguity, and any form of disorderly conduct.[4] Ironically, Abhor finds in the amoral figure of her father the impetus to defy the oversimplification of the world, embrace pain, and seek out complexity. Just as her grandmother, charged with prostitution, realizes that "the Vice-Squad swore whatever the Vice-Squad swears in order to maintain the scheme of things. Which might or might not exist" (5), so Abhor realizes that the patriarchy has imposed a falsely simplified order to maintain its power. Her father's transgression introduces her to a complexity in the world which is both more painful and more truthful.

Consistency in the narrative style of a novel orients the reader to the social order that the fiction describes. The sophistication and urbanity of the Jamesian narrator constantly reminds the reader of the Anglo-European aristocracy through which he moves. The folkloric and dialect quality of Samuel Clemens's *Adventures of Huckleberry Finn* locate the reader within a mid-American community and an unpretentious, populist milieu. Low-life confidence men and swindlers such as the king and the duke appropriate and mangle elevated discourse to comical effect: "I say orgies, not because it's the common term, because it ain't—obsequies bein' the common term—but because orgies is the right term. Obsequies ain't used in England no more, now—it's gone out. We say orgies now, in England. Orgies is better, because it means the thing you're after, more exact. It's a word that's made up out'n the Greek *orgo,* outside, open, abroad; and the Hebrew *jeesum,* to plant, cover up; hence in*ter*. So, you see, funeral orgies is an open er public funeral" (134). Acker disturbs the confidence with which the reader ascertains the social order from narrative discourse. Such transgression of the boundaries between high and low art forms is a recognized tenet of postmodernism. Acker's Abhor is supposedly illiterate because "being black, she was uneducated" (*Empire* 201). She generally speaks in the patois of the urban teenager. But her language and intellectual engagement just as often smack of the political theorist or the psychoanalyst. Acker herself has been a student of the urban punk movement, left-wing political theory, and post-Freudian analysis, but the turbulent mixing of these discourses does violence upon the readerly conventions of fiction. The first-person narration in which Abhor accounts for her father's amoral upbringing provides a good example of such cross-coded discourse:

> [H]e had no morals, for any morality presumes a society. Since my grandmother loved him, she saw no reason to teach him anything or that he should learn anything.
>
> This substitution of *primitivism* which must be *anarchic* (in its nonpolitical sense) for *morality* gave my father his charm. His charm blinded not only his parents but even every old farty schoolteacher to both his complete lack of social awareness and of education. Politics, for my father, was, always, a hole. (8)

Abhor's defense of her father's primitivism (which she supposedly shares) and her attack on the educatory system of "old farty schoolteachers" that enforces both a moral code and social order is nevertheless couched in the abstract terminology and the structure of argument that suggests a higher standard of education. Acker charges that education is chiefly a measure of the individual's training in obedience to the dominant culture. In the same gesture she defeats the readerly conventions that have trained us to expect a consistent narrative style and characters that remain locked within the referents of a single social class.

The scumbling of levels of discourse in the novel reflects Acker's anarchistic methodology, undermining the reader's presuppositions of dominant-intellectual and subordinate-proletarian cultural positions. Acker combines the use of essay-like titles for several subsections of the novel, such as "3. Beyond The Extinction of Human Life" (31), with crude and frankly obscene passages: "'If you finance her fucking for money,' said my father whose IQ was 166, 'I'll let her do it.' My father knew his mother-in-law was the cheapest thing on earth, even cheaper than himself" (16). Her characters are just as likely to be pimps as economics professors and she challenges the social order that distinguishes too finely between the two. Intensive, poetic, and gratuitous obscenity, "almost as beautiful as a strand of my grandmother's cunt hair" (4), can be followed without pause by the high abstractions of literary theory:

> The German Romantics had to destroy the same bastions as we do. Logocentrism and idealism, theology, all supports of the repressive society. Property's pillars. Reason which always homogenizes and reduces, represses and unifies phenomena or actuality into what can be perceived and so controlled. The subjects, us, are now stable and socializable. Rea-

> son is always in the service of the political and economic masters. It is here that literature strikes, at this base, where the concepts and actings of order impose themselves. (12)

This passage, no doubt plundered from her reading, appropriates theory in defense of Acker's radical poetics. Like the German Romantics or the anarchist who assassinated President William McKinley at the Pan-American Exhibition in Buffalo, NY in 1901, Acker intends to shock (through obscenity) and upbraid (in theory) the bourgeoisie. For Acker, the purpose of literature is indeed to assault the authority figures of an imposed order. The conjunction of obscenity and literary theory so that one is no less capable of affective impact than the other, and the very introduction of a passage of commentary in the midst of an uneducated character's discourse, work to deconstruct (or fuck up) the reader's well-trained expectations for novelistic discourse.

But why stop there? Acker commends Sylvère Lotringer for introducing her to the "French philosophes" Gilles Deleuze and Michel Foucault: "I didn't really understand why I refused to use linear narrative; why my sexual genders kept changing; why basically I am the most disoriented novelist that ever existed." Their theory places "this whole language at my disposal. . . . I know exactly what they're talking about. And I could go farther" ("Devoured" 10). In addition to the conflicting modes of discourse in *Empire of the Senseless*, Acker's disorientation as a novelist compels the reader to abandon other orderly conventions of reception. The many irruptions in causality in the novel defeat the reader's understanding of why one event proceeds from another. This apparent disorder collides with the vestiges of argumentative structure and the hectoring tone of the narration that propels the reader through the three sections of the novel. A temporal uncertainty ensues from the setting of the novel in a future-imperfect world (after the sack of Paris by the Algerians), yet with many references to the institutions and insults of the Reagan Eighties.

One section of "Elegy for the World of the Fathers" carries the academic subtitle "1. A Degenerating Language" (53). Acker's novel attempts a daring break from the prison-house of patriarchal language, shattering signs and conventions as she goes. At times she makes her text unreadable, "a sign of nothing" (53). She interpolates text in Persian, an invasion of the Western literary tradition by an Eastern literature which few are capable of reading except in translation. The Persian text, which is read from right to left, literally disrupts the flow of her text in English from left to right. Although the Persian text is "unknowable" for most readers, it still signifies the arbitrariness of literary conventions. Logocentrism is indeed supportive of a repressive society.

Acker views the degeneration of language, the collapse of the rules that guide and restrain the writer, as a largely positive effect of the death of the patriarchy. As a literary terrorist, she does her part in eroding standard English constructions and promoting slang expressions. The conditions of language, sexuality, and anarchy—the word, the body, and the body politic—are directly related: freedom in one sphere is an expression of freedom in the others; repression in one is signatory to repression in the others. This relation has a disturbing effect on daddy:

> Here language was degraded. As daddy plumbed and plummetted [*sic*] away from the institute of marriage more and more downward deeply into the demimonde of public fake sex, his speech turned from the usual neutral and acceptable journalese most normal humans use as a stylus mediocris into . . . His language went through an indoctrination of nothingness, for sexuality had no more value in his world, until his language no longer had sense. Lack of meaning appeared as linguistic degradation. (17; Acker's ellipsis)

Daddy's "stylus mediocris" remains in force so long as he is bound to the moral strictures and social institutions that restrict his personal freedom. Such "neutral and acceptable journalese" is a commodity of exchange sanctioned and controlled by the social order, permitting little if any individual expression. Personal expression becomes equated with deviance because the sense-value isn't immediately transmittable. The degradation of daddy's language and sexuality is expressed in terms of a pirated text riddled with lacunae: "This is what daddy said to me while he was fucking me: 'Tradicional estilo de p . . . argentino. Q . . . es e. mas j . . . de t . . . los e . . . dentro d. . . . ' He had become a Puerto Rican" (17; all but the final set of ellipses are Acker's). The unreadable Spanish text ironically codifies the qualities of traditional style in Latin America. Acker's anarchism acknowledges a lapse into "nothingness," the absence of rules and the disappearance of conventional values. Daddy's transgression of the taboo against incest consigns him to a subordinate and oppressed racial identity and to unintelligibility. But for Acker linguistic degradation is also linguistic freedom. Language confined to the transmission of codified values has no appeal. Language liberated from conventional meaning entertains at least the possibility of individual expression amidst disorder.

In order to understand the oedipal society, Acker devotes her attention to the Sadean obsession with dominance and punishment in the first part of *Empire of the Senseless*. As with her incestuous desires, Abhor experiences a strange attraction to the forces of control and obedience: "I saw a torturer. He was spending most of his time sticking electrodes on the genitals of men who protested against the government. He was paid to do what he was told to do because he had a job. I saw I wanted to be beaten up. I didn't understand" (51). A beating would confirm Abhor's sense of marginality in the oedipal world; pain insures her identity as renegade "other." But her confusion only reflects the role of sadism within patriarchal culture. Sade was imprisoned or confined to asylums for 27 years, punished for what society called the deviance of his devotion to sexual violence and physical discipline. His pleasure was his deviance, the violation of socially adjudicated rules and inhibitions. Michel Foucault points out that "in Sade, sex is without any norm or intrinsic rule that might be formulated from its own nature; but it is subject to the unrestricted law of a power which itself knows no other law but its own. . . . [T]his exercise carries it to a point where it is no longer anything but a unique and naked sovereignty: an unlimited right of all-powerful monstrosity" (*History* 149). For Sade the denial of inhibiting regulation leads to despotism. Abhor recognizes that the perennial devotees of sadistic punishment are those regimes most concerned with enforcing conformity to phallocentrism. The torturer operates within an economy of instruction; one is paid to do what one is told. Sadism is both attractive for its deviance and repulsive as a weapon beloved by right-wing dictatorships. Acker makes that connection explicit in her comments on *Empire:* she turned to "the Marquis de Sade because he shed so much light on our Western sexual politics that his name is still synonymous with an activity more appropriately named 'Reaganism.' Something of that sort" ("A Few Notes" 35). Sadomasochism, which derives sexual pleasure from simultaneously inflicting pain and enduring punishment, informs Abhor's reaction: "This isn't enough. Nothing is enough, only nothing. I want to get to what I don't know which is discipline. In other words I want to be mad, not senseless, but angry beyond memories and reason. I want to be mad" (51). Abhor's quest for anarchy, which she cannot know within the patriarchal system, leads her in the track of discipline. Although a half-robot, Abhor wishes to depart from the empire of the senseless where her behavior is automatized—performed unconsciously—in order to emerge among the "mad," in revolt against reason.[5]

In the revolution that closes Part I of the novel, Algerians led by a one-armed proletarian guerrilla named Mackandal wrest Paris from the control of François Mitterrand, the bourgeoisie, and the French government.[6] "Paris was in chaos. Thousands of Algerians were walking freely. Ragged. Dirty. Sticks. Dolls. Voodoo" (67). Acker considers the only remaining source of resistance to Western capitalism and its homogenized culture to be the Muslim world. "I thought, for Westerners today, for us, the other is now Muslim.

In my book, when the Algerians take over Paris, I have a society not defined by the oedipal taboo" ("A Few Notes" 35). The principal methods by which the French middle class retain their exploitative control of the African labor force are the media and disease. The Algerians chant "With this cry—MASTER—reap your profits in us, out of us. With this cry, by means of your press, press and oppress us" (70). The inescapable saturation of the media controls the minds of the masses, disciplining them to their social responsibilities and enslaving them to materialist needs. But in Acker's near-future world, disease controls the proletarian body: the Algerians protest, "No longer will you work in our muscles and our nerves creating herpes and AIDS, by doing so controlling all union, one and forever: being indivisible and narcissistic to the point of fascism, you have now closed down shop" (71). Sexually transmitted diseases become a device of the moralist middle class in suppression of the proletariat. Their "union" is corrupted, their bodies are stigmatized. Pervasive STDs are an intrinsic form of punishment that holds the Algerians in check. Mackandal's response is a terrorist campaign that involves the poisoning of the Parisian middle class using readily available herbs: "Poison entered the apartments of the bourgeoisie. There is a way to stop guns and bombs. There's no way to stop poison which runs like water. The whites had industrialized polluted the city for purposes of their economic profit to such an extent that even clean water was scarce. They had to have servants just to get them water and these servants, taught by Mackandal, put poison in the water" (77). As the bourgeoisie sought to discipline the proletariat through the transmission and stigmatization of a naturally occurring disease, the revolution of the dispossessed is fostered by the introduction of a natural toxin. The terrorism succeeds by striking at the weaknesses created by the bourgeoisie's exploitation of their human and natural resources. Disease as discipline confronts its twin in poison as anarchic revolt.

The Dominion of Anarchy: Breaking the Code

The second part of the novel, "Alone," explores the possibility of society in which phallocentric domination "on the political, economic, social, and personal levels" has collapsed ("A Few Notes" 35). The section is preceded by a tattoo-drawing of a storm-buffeted schooner with sails furled, pitching through the roiling waves. The legend below reads "The Deep" (88). Acker's figure for individual freedom from social restraint is the pirate. "Sailors leave anarchy in their drunken wakes" (113). But the turbulence that follows in the wake of these (now frequently female) pirates can be as creative as it is destructive. In a chapter titled "The Beginning of Criminality / The Beginning of Morning," Acker advances an aesthetic and political theory of creative disruption. Criminality as a violation of the established law is regarded as destructive; but criminality in an environment of instability has the capacity for genesis—the beginning of morning. During the dark centuries of European colonization and enslavement of Africa, the Algerian coast became a stronghold of pirates. The oppressive rule and the claim to human property by Europeans fosters "criminal" transgression of property law. In the Algerian revolution (1954–1962), those Muslims who fought the war for independence to dislodge the French from Africa turned criminal revolt into nation-building. Abhor proposes that "All good sailors espouse and live in the material simplicity which denies the poverty of the heart. Reagan's heart is empty. A sailor is a human who has traded poverty for the riches of imaginative reality" (114). Acker's pirates are also equated with artists whose creative powers enable them to transcend materiality. "Such an act constitutes destruction of society thus is criminal. Criminal, continuously fleeing, homeless, despising property, unstable like the weather, the sailor will wreck any earth-bound life" (114). The artist, the sailor, and the revolutionary challenge the Western illusion of material permanence and ownership. Their production of the new arises only out of the destruction of the old: "Though the sailor longs for a home, her or his real love is change. Stability in change, change in stability occurs only imaginarily. No roses grow on sailors' graves" (114). Acker proposes a generative instability.

Acker challenges the literalist who perceives order only in stable forms and unchanging institutions. These literalists are invariably shocked when confronted by the sudden obsolescence of a familiar order because they have denied the process of change. They cannot understand that *an* order is always impermanent, and as a limiting and limited case, not to be taken as *the* order of things. It emerges from a vast array of possible orders—as much by accident as by action. Like a wave it eventually dissipates, possibly supplanted by another formation. One proponent of "stability in change" is the male pirate, Thivai. While searching for a whore, he declares, "If there is any variability to reality—functions which cannot be both exactly and simultaneously measured—reality must simultaneously be ordered and chaotic or simultaneously knowable and unknowable by humans" (102–103). Just as the "beginning of criminality" (or the violation of an apparent order) can also be the "beginning of morning" (or the creation of a new system), so the presence of variability in reality signifies the simultaneity of order and chaos. It takes a pirate mind to appreciate that order and chaos, the measurable and the unmeasurable, are not exclusive to one another in experience but co-present, the one emanating from the other in continuous change and exchange.

Acker's pirates question whether "the demand for an adequate mode of expression is senseless" (113). The pursuit of an adequate means of expression leads to conventional discourse, its methods and rules governed by the empire. The unique expression of the self is imprisoned, rendered senseless to the individual in being made accessible to and consumable by the many. Acker challenges the function of adequate language through the tattoo as writing and the flagrant disregard for convention and decorum in a taboo language. For Acker, "writing the body" represents a profoundly ambiguous system of signification and is thus less susceptible to commodification and devaluation by the empire.[7] The arrival of Abhor and her newly acquired friend, Agone (a male Cuban sailor) at a tattoo parlor allows for a short disquisition on the etymology and cultural history of the art form:

> Cruel Romans had used tattoos to mark and identify mercenaries, slaves, criminals, and heretics.
>
> For the first time, the sailor felt he had sailed home.
>
> Among the early Christians, tattoos, stigmata indicating exile, which at first had been forced on their flesh, finally actually served to enforce their group solidarity. The Christians began voluntarily to acquire these indications of tribal identity. Tattooing continued to have ambiguous social value; today a tattoo is considered both a defamatory brand and a symbol of a tribe or of a dream. (130)

The tattoo can be both an artistic expression and an identification mark. It is an embellishment of the body through design, and an ineradicable injection of poison ink under the skin. The Romans used the tattoo on the early Christians, and the Nazis on European Jews, to identify and control the heretic and pariah. The sailor welcomes this identifying mark of difference. He celebrates his outcast status. That writing which facilitated the control and purging of the undesirable becomes an expression of group identity. The double value of "defamatory brand" and "symbol of a tribe," purging and bonding through identification of difference and sameness, a painfully forced marking and a defiant self-expression makes the tattoo an ideal signifying system for Acker. In searching for a myth in *Empire of the Senseless*, Acker says that the "most positive thing in the book is the tattoo. It concerns taking over, doing your own sign-making. In England . . . the tattoo is very much a sign of a certain class and certain people, a part of society that sees itself as outcast, and shows it. For me tattooing is very profound. The meeting of body and, well, the spirit—it's a *real* kind of art, it's on the skin" ("Conversation" 17–18). The Algerians, pirates, and sexual pariahs in the novel seize control by taking as their own the very means by which the empire has suppressed them. The tattoo is a figure for Acker's art: a non-linear writing that foils the plot-driven causality of traditional fiction; an all-over writing that entwines its subjects rather than pursues its conclusion; a writing of the body (either female or male) expressive of sexuality and the visceral as opposed to the rational, concentrated mind. The tattoo-drawings that appear in the novel—at the start of each section, and at the conclusion—are more than illustrations: they are expressions of the artist's control of her medium, signs of her revolutionary discourse.

The disquisition on tattoos in the novel concludes with the declaration, "In Tahitian, writing is 'ta-tau'; the Tahitians write directly on human flesh" (130). On the equation of writing and tattooing Acker says, "I'm fascinated with the relationship between language and body. . . . I'm interested in the material aspect of the tattoo. . . . Erotic texts at their best—I don't mean pornographic, which is something else—are very close to the body; they're following desire. That's not always true of the writer, whereas it's always true that the tattooist has to follow the body. That's the medium of tattoo" ("Conversation" 18). Acker's novels follow desire, and they seek a "language of the unconscious." Intervening in the action as the tattooer approaches Agone with a knife, Acker calmly describes the crisis in language: "That part of our being (mentality, feeling, physicality) which is free of all control let's call our 'unconscious.' Since it's free of control, it's our only defence against institutionalized meaning, institutionalized language, control, fixation, judgement, prison" (133–34). Acker wishes to emulate the freedom of the unconscious in her writing, to pursue desire without arresting it, to wrest language from its civilizing discipline and reinvest it with an anarchic impulse.

Acker reflects on her efforts to free language from institutional control: "Ten years ago its seemed possible to destroy language through language: to destroy language which normalizes and controls by cutting that language. Nonsense would attack the empire-making (empirical) empire of language, the prisons of meaning" (134). The cut-up texts of William Burroughs, for instance, employ aleatory operations to defeat the writer's culturally infused determinations and thus oppose the empire's senseless blague with nonsense. To some degree Acker's plagiarism as an excision of classic texts participates in this conceptualism. "But this nonsense," she recognizes, "since it depended on sense, simply pointed back to the normalizing institutions." The binary opposition imperils avant-garde writing as simply the negation of conventional discourse, defined by its opposition to the institutional code without actually eliminating those codes. Acker prefers instead to attack the codes themselves through speech that is not unintelligible but forbidden, to speak the unspeakable:

> What is the language of the "unconscious"? (If this ideal unconscious or freedom doesn't exist: pretend that it does, use fiction, for the sake of survival, all of our survival.) Its primary language must be taboo, all that is forbidden. Thus, an attack on the institutions of prison via language would demand the use of a language or languages which aren't acceptable, which are forbidden. Language, on one level, constitutes a set of codes and social and historical agreements. Nonsense doesn't per se break down the codes; speaking precisely that which the codes forbid breaks the codes. (134)[8]

The prison houses and asylums are suffused with the shouts and screams of their inmates. The system has no need to suppress the nonsensical protests of the damned and convicted. But these institutions forbid the subversive communication that passes secretly from cell to cell—plotting the demise of the guardians of civilization and the downfall of the wardens of culture. Punishment is the silencing and the isolation of the lawbreaker. Acker smashes codes and disrupts societal agreements through the insinuations and assaults of proscribed speech. Her achievement lies not in opposing the linguistic code with meaningless cipher but in forcing the repeal of prohibitions through relentless violation—the code must then be revised to permit what it previously denied.

Plagiarism is the fundamental taboo of the literary world, and Acker has paid a price for violating its ethic. If Hannibal Lecter is her father, she has cannibalized the bloated corpse of patriarchal fiction. In her conversation with Lotringer, she recounts:

> A collection of my earliest work was published in England, and in the section about TOULOUSE LAUTREC there's four pages which I took out of a Harold Robbins novel, a best-selling book called *The Pirate* that had been published some years before. There is a scene there where a rich white woman walks into a disco and picks up a black boy and has sex with him. I changed it to be about Jacqueline Onassis and I entitled the piece "I Want to Be Raped Every Night. Story of a Rich Woman." I think the joke's quite obvious, but the journalist called my publisher and then she called Harold Robbins' publisher, and their response was that, my God, we've got a plagiarist in our midst. So they made a deal that my book would be immediately withdrawn from publication and

> that I would sign a public apology to Harold Robbins for what I had done. ("Devoured" 12)

Acker had no intention of deceiving her readers in the appropriation of Robbins' novel. She's "always talked about [plagiarism] as a literary theory and as a literary method" ("Devoured" 13). Even so, she has *pirated* Robbins's *The Pirate,* seizing its precious cargo and redirecting it for her own purposes. "Hijacking a copyright, no wonder they got upset," says Lotringer, "Terrorism in literature" ("Devoured" 13). To deceive a reader by appropriating a work as one's own would only reinforce the taboo against plagiarism and strengthen the capital investment in copyright.[9] Changing the intentionality of Robbins's "soft core porn" by intensifying the sexuality of the language, Acker breaks the code—of a patriarchal fondling of women's bodies, of protection of literary property—by speaking precisely that which the code forbids.

Another literary code that Acker flagrantly violates is that of genre. She upsets the literary hierarchy of genres that designates the linguistic difficulty of poetry or the subtle aestheticism of the *künstlerroman* as elite modes and the world-making of science fiction, the action-adventure novel, and the sexual arousal of pornography as populist modes. Her method in this postmodern leveling of high and low cultural signifiers is the mixing of familiar characteristics of these genres when the literary code would normally demand their separation. Leslie Dick points out that genres acquire their power through the formulaic assertion of rules: "a specific genre can be an 'enabling device,' a formal structure that allows and controls and prevents meaning, a syntax" (208). Acker frequently invokes the syntax of genres only to thrust the pelvis of a taboo language at them. The pirates' quest ends in pornographic satisfaction, as the obscene reality of buccaneering confronts *Treasure Island:* "the pirate who had just been fucked bent over the child tightly bound in ropes, already raped. His hands reached for her breasts. While sperm which resembled mutilated oysters dropped out of his asshole, he touched the breasts" (*Empire* 21). The figurative language only further confounds genre rules. In response to the rigidity of literary genres, the artist can (according to Leslie Dick) "isolate and extract forms from the institution, without getting involved in the institution" (209). Acker's piracy certainly extracts cultural booty from high and low genres, but her aesthetic of the taboo demands a frontal assault on the institution; or as Dick concludes, "it's more like scavenging, ripping the genre off. It's making use of some of the elements of the genre, while discarding the implicit values of the genre as institution. It's destructive, and disrespectful of the genre, which it treats like an abandoned car" (209). As an example, Acker introduces the character of

Sinbad the Sailor among her rogues in the second part of *Empire*. Although lifted from the oriental tale, he describes his family history in the code of science fiction: "My father, I remembered, came from Alpha-Centauri. His head, the case with most Centaurians, had been green and flea—or dried-drool—shaped. Unlike him, my mother, a moon-child, was just a good-for-nothing. She was beautiful by night-time or lightless standards. Like the moon which hides behind the sun, mommy kept her brains hidden" (154). This passage takes advantage of the science-fiction genre's capacity for invention—any combination is possible in a new world—but still attacks the present code of gender inequity which relegates female intelligence to pale reflection of the dominant bull-Centaur. Sinbad reflects that he is thus a product of "cross-racial union. Multi-racial marriages usually lead to disaster" (154). In crossing the codes of genres, Acker challenges the institutional enforcement of racial and literary homogeneity. Her multi-generic fiction is one method by which the institutional code of language is broken.

Her Rules: Re-Inventing the Code

In the third part of *Empire of the Senseless*, "Pirate Night," Acker's renegades are questing after a "society that is taboo," that is established on the very ground of transgressive acts. This section begins by plagiarizing *Huckleberry Finn*, which Acker describes as "one of the main texts about freedom in American culture" ("A Few Notes" 36). Thivai and his gay friend Mark (Huck and Tom) go to elaborate lengths to liberate the imprisoned Abhor (Jim), who is part black and referred to as a "runaway nigger" (*Empire* 212). In Acker's transformation, Abhor's enslavement is not to racism but to sexism; after much effort to spring her from prison, Abhor walks out of the jailhouse unmolested and determined to form a motorcycle gang. Male assistance in breaking the chains of sexism is worse than ineffectual; on the other hand, breaking out of the restraints of the old order won't suffice to establish a new society. Abhor forms a motorcycle gang "because motorcycle gangs don't let women ride bikes" (212). Rather than protest her exclusion from a male-dominated activity or attempt to dissolve the institution for its discrimination, she acts on her own to establish a social order based on that tabooed behavior. In one of many ruptures of causality and probability in the novel, Abhor—lacking the cash for a new or used bike—simply "finds" the motorcycle, gasoline, and oil that she needs: "I turned around, walked into the woods, and found a Honda which was only a year old, prerevolutionary, and in perfect shape except for one cracked mirror" (211). Abhor's unlikely discovery occurs outside of the capitalist economy of labor, wages, and the purchase of commodities; and it also violates the fictional economy of motivation that demands a reason for the motorcycle's presence when and where it's desired. Launched into a taboo society, the forbidden is not only permissible but unquestioned.

Abhor's motorcycle adventures represent the struggles involved in the creation of a taboo society. Never having been allowed to ride a motorcycle, Abhor becomes frustrated as she attempts to engage the motorcycle's clutch properly: "I got angry at the clutch and called him or her a shitsucker. This showed that both men and women do evil. But this knowledge and understanding didn't help me deal with my clutch" (213). Acker toys with the politically correct injunction against sexist language, in particular those instances when the speaker adopts the masculine pronoun to refer to someone whose sex cannot be identified. As a recently published handbook for writers advises, "If you want to avoid sexist language in your writing, follow the guidelines in Chart 99. Also, you can avoid sexism by avoiding demeaning, outdated stereotypes, such as *women are bad drivers* or *men are bad cooks*" (Troika 400). Abhor hopes to shatter the stereotype that a woman is incapable of riding a motorcycle. But the clutch—which obviously has no gender as an inanimate object—frustrates her efforts. Both genders are capable of becoming angry and resorting to profanity. Following the non-discriminatory principle of non-sexist language, Acker (through the persona of Abhor) admits that her taboo society needn't be theorized as a feminist utopia in which violence and evil have been completely expunged. The world of "Pirate Night" puts the brakes on such feminist fantasias as Abhor concerns herself with the challenges of vehicular realpolitik. Abhor learns that "a clutch controls power; to get more power, you have to control power. That was good" (213). Only recently freed from patriarchal captivity, Abhor learns to master power on her own, for her own purposes. She refuses to identify evil with a single class of oppressor, nor does she deny the continued need for discipline and control.

Abhor's male accomplices are skeptical, however, of her ability to ride because she doesn't know the "rules of road behaviour. They're found in a book called *The Highway Code*" (213). Her demonstration of individual capacity and self-governance runs afoul of this instance of "prerevolutionary" regulation and restriction of freedom. Abhor declares, "I had never heard of any rules so I didn't know that there were any, so I went back into the woods where I found a wet copy of *The Highway Code*. This was an English book, dated 1986. I had the CODE so now I could drive" (213). In *Elements of Semiology* Roland Barthes points out that the Highway Code is one of the most intractable of semiological systems: "by reason of its very purpose, which is the immediate and unambiguous understanding of a small number of signs, the Highway Code cannot tolerate any neutralization" (84). Neutralization refers to "the phenomenon whereby a relevant opposition loses its relevance, that is, ceases to be significant" (83). Thus the opposition of red and green as signifiers and their signified behavior must be maintained. The Highway Code is irreproachable. Or as Barthes suggests, it "must be immediately and unambiguously legible if it is to prevent accidents" (80). Barthes describes the Highway Code as one of the most rigid and limiting of semiological systems because, unlike the fashion system or literature, it forbids polysemy. It is a distinctly masculine, authoritarian system that regulates by establishing arbitrary but inflexible conventions. Abhor learns to ride by feel and intuition; the masculine code is a deterrent to her feminine experience.[10] She observes, "Its first rule for bikers said that a biker should keep his (I had to substitute *her* here, but I didn't think that changed its sense) bike in good condition. Since this bike wasn't mine, I could keep her in any condition. Since this is only commonsense and commonsense is in my head, I tore out this section of *The Highway Code* and tossed it into a ditch" (213). The masculine Code's presumptions of appropriate gender behavior, its discriminatory language, and its property fetishes don't deter Abhor from being a freethinker or acting on empirical observation. Nevertheless, her fitful attempts to follow the Code rather than discard it entirely lead her to vehicular chaos. The restrictions of the Code in a postrevolutionary era provoke anarchic behavior.

No theme is more prevalent in issues of *Outlaw Biker* magazine than the libertarian beckonings of the open road. And yet Abhor halts at the intersection between transgressive-taboo behavior that acknowledges the Code in its violation of it and a fully invested, self-determined behavior that invents its own customs. In order to disrupt the male Code of the Motorcycle Gang that forbids the woman driver, Abhor must first be initiated into the Rules of the Road. She pays literal attention to

> Rule 55.
> c) Watch your speed; you may be going faster than you think.

She assumes that she "was driving correctly by staring down at the speedometer" (218). Following the rules, when the rules don't mediate female experience, has its dangerous consequences; the result is a rear-end collision with a truck. Following the code prevents following the road. As the injured truck driver approaches menacingly, Abhor realizes that one paradigm of behavior has been destroyed by the revolution though no other has yet taken its place: "I was confused about what was happening because there were no more rules.

Perhaps I was on the crossroads of Voodoo" (218). The remnants of patriarchal order and its unambiguous Code are thus incompetent in the world of orderly disorder into which Abhor has rushed and from which a new Code has yet to be created. The monologic, hermeneutically forbidding Highway Code has given way to the terrible beauty of a free polysemy. Abhor has entered the domain of crossed signals and Voodoo intersections: "One road was that the old man was trying to give me an important message. The other road was that the old man was trying to kill me" (218–19). Rather than the neutralization of opposing signals, Abhor encounters the ambiguously legible, potentially fatal, crossing of the blinking yellow light.

In denouncing the impositions of patriarchal order, Abhor (most probably speaking for Acker) issues the politically charged pronouncement that "the problem with following rules is that, if you follow rules, you don't follow yourself. Therefore, rules prevent, dement, and even kill the people who follow them. To ride a dangerous machine, or an animal or human, by following rules, is suicidal. Disobeying rules is the same as following rules cause it's necessary to listen to your own heart" (219).[11] Acker endorses an essential principle of anarchism, that the impulse to personal freedom more frequently leads to salutary and creative behavior, whereas authoritarian strictures imposed on individuals more frequently lead to violent and destructive behavior. In their essay on the discourse of the Paris Commune of 1871, Donald Bruce and Terry Butler point out that anarchy has been historically "synonymous with the notion of chaos understood to mean 'utter confusion, the absence of all order, disorder.'" In the Paris Commune, however, and in the postrevolutionary Algerian-controlled Paris of Acker's *Empire*, the "actual political notion . . . should rightly be understood as the *radical decentralization of authority:* in other words a type of *order within disorder* (231, their emphasis). Abhor's repudiation of externally imposed authority in favor of listening to the heart implies a commitment to spontaneity, self-discipline, and self-organization. "From now on *The Highway Code* no longer mattered. I was making up the rules" (*Empire* 222). The anarchy of Acker's postrevolutionary Paris conforms most closely to the definition of chaos that provides for an intrinsically originating creative dimension, that is, as an "*order [which] arises out of chaotic systems*."[12] Drawing here on the propositions of Ilya Prigogine and Isabelle Stengers in *Order Out of Chaos*, Bruce and Butler argue that in a community in which self-organization and spontaneity are emphasized, "an explanation of chaos envisages (physical and social) systems which are capable of renewing themselves. Instead of falling victim to the inescapable entropic diffusion of all energy concentrations in the physical and social universes (as postulated in the second law of thermodynamics), this notion of chaos proposes a theory of renewal by which complexity arises out of simpler physical and social systems as a response to surrounding conditions" (235–36). In the chaotic social system of "Pirate Night," Acker theorizes a generative anarchism that initially hastens the collapse of repressive systems characterized by a simple homogeneous order, and in the erosion of closed-system boundaries, has the negentropic capacity to invigorate the newly opened environment. Abhor's declaration that "I was making up the rules" represents the possibility of self-organization and renewed complexity in an anarchic, heterogeneous society. Acker's confidence in the creative capacity of anarchy invokes a comparison with the poet Arthur Rimbaud,[13] whose call for a "'dérèglement de tous les sens' constitutes a chaotic subversion of order the aim of which is to let *emerge* a spontaneous and as-yet-unknown order" (Bruce and Butler 236, their emphasis). Abhor's world suggests as well a "twofold vibration" (to borrow the title of Raymond Federman's novel) comprised of sadism and masochism, the instigation of anarchism and the declaration of self-governance, spontaneity and organization. She reflects, "I'm the piercer and the pierced. Then I thought about all that had happened to me, my life, and all that was going to happen to me, the future: chance and my endurance. Discipline creates endurance" (224). A new order emerges from the subversion of the old; anarchy and discipline are closely entwined. Anarchy permits the irruption of chance that may destroy life, but discipline fosters the endurance that sustains life.

Abhor's creativity takes the form of rewriting the authoritarian *Code* as a book called *The Arabian Steeds* because "My heart said these words. Whatever my heart now said was absolutely true" (219). Purportedly illiterate, Abhor draws pictographic images over the familiar diamond-shaped warning signs of the *Code* that are reproduced in the novel, icons of a Western industrialized, petroleum-dependent, contaminated, asphalt-topped, inflammable society. Abhor converts, deciphers, and performs a hermeneutical transformation of the warning signs of an industrialized world in collapse into the vitalism, free will, and Bedouin-nomadic values of North African cavaliers: the motorcycle becomes Arabian steed; the partitioned and industrialized city of Paris becomes open desert; cold metal becomes hot sand; and masculine becomes feminine. Thus, for example, the icon for "corrosive" (a property of acids that causes the gradual destruction of metals, and so inimical to an industrialized society) is translated as "Let anger be anger: neither self-hatred nor self-infliction. Let the anger of the Arabian steeds be changed through that beauty which is blood into beauty" (221). In Acker's conception of the post-revolutionary nomadic social order anger isn't neutralized—as the chemist treats an acid with a base—but transformed in a salutary, sublimated manner. The corrosive aspect of anger, as a warning sign of the entropic breakdown in the steel sinews of industrialized Western culture, becomes a proclamation against a psychological and emotional breakdown in the individual through self-hatred. Anger is transformed, not neutralized in this post-patriarchal society from self-infliction to self-definition. Abhor's adoption of the nomadic creed expresses the release or de-institutionalization of the individual from the metallic cocoon of Western culture. Here Acker appears to draw upon concepts of "deterritorialization" and "smooth space" in the work of Gilles Deleuze and Félix Guattari, especially in their discussion of the nomad: "The nomads are there, on the land, wherever there forms a smooth space that gnaws, and tends to grow, in all directions. The nomads inhabit these places; they remain in them, and they themselves make them grow, for it has been established that the nomads make the desert no less than they are made by it. They are vectors of deterritorialization. They add desert to desert, steppe to steppe, by a series of local operations whose orientation and direction endlessly vary. . . . The variability, the polyvocality of directions is an essential feature of smooth spaces of the rhizome type" (382). The free range, self-definition, and polyvocality of nomadic life express Abhor's liberation from and transformation of the fixities, entropophobia, and hortatory univocality of Western society.[14]

With something of the air of a cartomancer turning over the last card at a reading, Abhor declares, "I drew a final picture which summed up all the other pictures" (*Empire*, 221). This picture appears in the text as the tattoo design of the rose-piercing sword, around which is the legend proclaiming *Discipline and Anarchy*. The enlacing banner does in fact summarize the interdependent domains of Acker's novel: sexuality and language; political identity and individualism; sadism and masochism; order and disorder; striated and smooth space; literary and subliterary genres; masculinity and femininity. Acker's tattoo of the piercer and the pierced entwined slashes at the designation of one component or subject position as the dominant and the other as the subordinate. She refuses to replace the crumbling patriarchal order in an oppositional hierarchy with a feminist-utopian world that merely reverses the polarity of values. Acker finds that even in the domain of anarchy —in nomadic space, after the disruption of the state apparatus, where women ride motorcycles—there must be discipline present. Just as Abhor attains her feminine identity through the realization of her anger, and in the chaos of the postrevolutionary state a self-governing system reveals itself, so discipline and anarchy are recognized as interdependent functions. Discipline without anarchy is repressive; discipline in anarchy promotes endurance. Anarchy without discipline is destructive; anarchy in discipline promotes creativity. Spontaneity *and* organization. Beauty *and* violence. A rose and a sword.

4

American Oulipo

Proceduralism in the Novels of Gilbert Sorrentino, Harry Mathews, and John Barth

> There is an abundance of incredible systems of pleasing design or sensational type. The metaphysicians of Tlön do not seek for the truth or even for verisimilitude, but rather for the astounding. They judge that metaphysics is a branch of fantastic literature. They know that a system is nothing more than the subordination of all aspects of the universe to any one such aspect.
>
> —Jorge Luis Borges, "Tlön, Uqbar, Orbis Tertius"

Partly by way of tweaking American pragmatism, and partly in admiration of the mechanical-industrial invention that made possible the modern city, the Dadaist Marcel Duchamp, transplanted from Paris to New York, quipped that "The only works of art America has given are her plumbing and her bridges."[1] As witty as it is French, the remark makes a truthful assessment of American tastes in the twentieth century, tending away from the decorative and favoring the expression of design. Recognition of the well-designed feature or product has been considered central to American commerce and culture, in most regards. The assembly-line manufacturing of Henry Ford's River Rouge plant in Dearborn, Michigan, the Art Deco features of the Chrysler Building in Manhattan, and the stolid American Standard fixtures that stand as glistening monuments in public buildings around the country testify that while Americans are indifferent to art, they appreciate design.[2] It would not be an exaggeration to claim that Americans consider aesthetic value only as it is comprised by the functionality of the object; what works well, performs to specification, is *ipso facto,* the beautiful thing. Thus it comes as something of a surprise that, when design is so prevalent in American industrialism, architecture, appliances, and furnishings, the explicit acknowledgment of design in American fiction is considered an affront to the common reader and a positive annoyance to the literary critic. The novel may be well-made, so long as it doesn't wear its signs of construction openly. This resistance arises not from evidence of functionality—from the demonstration of literary technique or style—in the work of fiction, but from a distaste for the apparent preconception or formulation of a plan, that the novel has been somehow "contrived" by its author. The anti-intellectual response to this sense of design is equally an American trait.

I have designated the adoption of a rigorous and efficient design in advance of the composition of a work of literature as proceduralism. It is important to distinguish this term from other critical analyses of structure in literature. Formalism establishes the characteristics of a text inherited within a literary tradition. The author who practices within that tradition relinquishes considerable control over the measure of these formal traits—the acceptable rhyming patterns of the sonnet, or the manipulations of a novel's plot that remain within the parameters of verisimilitude—and, in the estimation of critics, demonstrates mastery to the extent that the deployment of these structural devices appears "natural" rather than conspicuous. Formalist analysis grants its most generous approval to those literary texts whose traits are commonly recognized among its type. This form of writing within a paradigm produces a text, for example the ratiocinative detective story in the manner of E. A. Poe or Arthur Conan Doyle, whose characteristics are largely determined for the author. Subsequent deviation from the formula demonstrates only artistic ineptitude. Proceduralists, however, invent forms without knowing the precise manner of text that will be generated. They welcome a degree of indeterminacy in literary production and do not confuse the value of the text with its conformity. The procedural parameters will often be unique to a given work. The prefatory commitment to design (often derived from extra-literary models) differs from formalism because the latter tests the skill or technique of the author in executing the literary model while the former risks all on true invention. Formalism is the equivalent of determinism in the physical sciences; proceduralism recognizes an essential uncertainty in the universe of its making. It's no wonder that procedural writing upsets the expectations of an American reader inured to popular genres and flaunts the critic's knowledge of the familiar characteristics of such literary works.

Literary proceduralism appears in both poetry and prose, in historical periods as distant as the classical era in Greek and Latin that gave us the palindrome and the acrostic, and in the Western and non-Western traditions alike. My attentions will necessarily be limited to a discussion of the contributions of American postmodern fiction. In particular, Gilbert Sorrentino's trilogy *Pack of Lies* (1997), Harry Mathews's *The Journalist* (1994), and John Barth's *The Last Voyage of Somebody the Sailor* (1991) are novels that reveal an immanent design within their apparently chaotic distribution of materials. Each book is constructed according to a set of arbitrary and precise rules, arrived at in advance by their author, with narrative consequences that are both generated and controlled by those rules. In the cases of Mathews and Sorrentino there is something of a French connection to the Paris-based Ouvroir de Littérature Potentielle (Workshop of Potential Literature), or Oulipo, founded in 1960 by François Le Lionnais and Raymond Queneau to investigate the possible uses of rule-based constraints in the composition of literature. Its members included Georges Perec, Jacques Roubaud, Jacques Bens, Marcel Bénabou, Italo Calvino, and Mathews (its only American inductee). According to Mathews, Le Lionnais and Queneau were both distinguished mathematicians, and Le Lionnais was one of France's leading authorities on chess problems (a subject that preoccupied Marcel Duchamp as well). The Oulipo thus had "a mathematical orientation that is fundamental, one that exerts an influence even on those members who are (to their regret) innocent of advanced mathematics" (Mathews, "Oulipo" 68). As a result several of the more well-known Oulipian experiments involve mathematical theorems, such as Queneau's combinatory sonnet, *Cent Mille Milliards de poémes* (one hundred thousand billion poems), analyses of permutational plot structures by Calvino and others, and the use of algorithms such as the S + 7 method (replacing each substantive noun in a text with the seventh following it in a given dictionary).[3] In Oulipian fiction the rule-based constraints usually have their greatest effect on what I call compositional method. Perec's novel, *La Disparition* (1969; translated by Gilbert Adair as *A Void,* 1994), is an example of the lipogram (an essay on the history of which Perec has written), a text in which a given letter of the alphabet is prohibited. Perec's avoidance of the normally ubiquitous letter *e* in French compels the author to dramatically reshape the prose idiom of his narrative, in this case naturally concerning a mysterious disappearance. Constraints such as the lipogram are thus syntactic or strategic. The constraint, as Marcel Bénabou explains, "forces the system out of its routine functioning, thereby compelling it to reveal its hidden resources" (41). Like a survivalist in unfamiliar terrain, the author under constraint must relinquish behavior that under normal conditions would suffice, and turn to unusual strategies for replenishment. Queneau's *Exercices de style* (1947; translated by Barbara Wright as *Exercises in Style,* 1981) recounts the same brief altercation between two men on a public bus in ninety-nine chapters, each written in a different style. This demonstration of stylistic virtuosity relies on the unpromising subject matter of a minor dispute to focus our attention and admiration on the performance of the author, that is, on the writing itself. In my introduction to this book I mentioned the closely related work of

Swiss-born Christine Brooke-Rose, who restricts the composition of her novel *Amalgamemnon* (1984) to the non-realized verb tenses (future, subjunctive, or interrogatory) and so compels her work into a speculative mode. Not an official member of the Oulipo, Brooke-Rose lives in France and taught at the University of Paris, Vincennes, from 1968 to 1988. By no means a complete catalog of Oulipian method, the restraints in these novels primarily affect the mode of composition. Although the restrictions govern the entire work, it is the syntax—or the material structure of the language—that is most noticeably altered.

The American proceduralists, whether they have been directly influenced by Oulipian practice or not, emphasize preconceptions of global structure in the novel, or what I have been calling design. These formal constructions, while not as evident at the level of syntax, can likewise serve as a constraint in the novel. Sorrentino admits that he adopts "a structure or series of structures that can, if one is lucky enough, generate 'content,' or, if you please, the wholeness of the work itself. Almost all of my books are written under the influence of some sort of preconceived constraint or set of rules. Some of these are loose and flexible, like the time scheme in *Steelwork* [1970], and others are quite rigorous, like the alphabetical framework in *Misterioso* [1989]. My interest in Oulipo dates back maybe 10 years or so, when I became aware of what they were up to" ("An Artist" 95). I'll return shortly to discuss the importance of the generative constraint in proceduralism, but first I want to attend to the global structures in these works. Sorrentino begins each novel in his trilogy, *Pack of Lies*, by settling on a structure or formal invention; the trilogy recapitulates the design of its individual parts. The structure that the author determines predicates the "wholeness" of each work. Thus the "alphabetical framework" of the third novel, *Misterioso*, should be complete in twenty-six sections (I'll return later to whether it is or isn't). Sorrentino proceeds with the conviction that "form determines content," and that these simple rules are capable of generating complex narratives.

In other procedural fiction, the use of a predetermined design can be represented in the form of a chart or graphical aid, which the novelist thoughtfully provides as an appendix to the work. In composing *The Castle of Crossed Destinies*, Calvino arranges the seventy-eight cards of the tarot deck into two intersecting designs that generate the narrative of the book's twin novellas, "The Castle" and "The Tavern." One of Calvino's self-imposed constraints is that all of the cards in the traditional deck—the twenty-two Major Arcana and the fifty-six Cups, Coins, Clubs, and Swords—be dealt in the form of a cross-hatched pattern whose tales are then "read" by the author in each direction. The two designs (among the many possible arrangements that Calvino considers) are represented as full-page illustrations in the book, in the Visconti-Bembo tarot for "The Castle" and in the Marseilles tarot for "The Tavern," and then in segments in the margins throughout the text.[4] It's worth noting that Sorrentino uses the Rider tarot to organize the seventy-eight chapters of his novel *Crystal Vision* (1981), though no illustrations are provided for the reader.[5] Two other works that provide graphical instructions for the reader are Perec's *La Vie, mode d'emploi* (1978; translated by David Bellos as *Life, A User's Manual*, 1987), and Barth's *LETTERS* (1979). The directions for reading Perec's novel appear as a full-page illustration immediately following the text in the form of a stylized rendition of the apartment bloc and its occupants at 11, rue Simon-Crubellier in Paris. As David Bellos has shown, a 10 × 10 grid of squares can be superimposed on this chart of residents. The ninety-nine stories (and an epilogue) follow the knight's tour problem in chess, so that every square in the grid is touched on one time only. The chart thus represents the design of the novel, its procedural order, and the wholeness to which the form aspires.[6] Barth employs the calendar for the seven months between March and September, 1969 on which he graphs the alphabetical characters and epistolary missives of *LETTERS*. The extended subtitle, "An Old Time Epistolary Novel by Seven Fictitious Drolls & Dreamers Each Of Which Imagines Himself Factual," is arranged to spell out the seven letters of the title each to a calendar month, their placement setting the order of exchange and the dates of dispatch between the seven correspondents (*LETTERS* 769). Barth speaks on behalf of the "elegant construction" of the novel in the foreword to the Dalkey Archive Press edition (*LETTERS* xvii). So large an undertaking (772 pages) benefits from an organizational superstructure, the reader appreciates the periodic guideposts, and the author is motivated by the preset order of letters. In general, American proceduralists adopt a variety of extraliterary forms to structure their works, whereas the Oulipians have been more concerned with syntactic or algorithmic devices.

The proceduralists who provide graphical designs for their novels are conspicuously advertising the constructed nature of their work. Not only does the fiction proceed according to a predetermined order, but the reader is actively involved, as it were, in the assemblage of the work. Why not, the author might ask of the reader, join me in cross-checking the progress of our enterprise, or take note of our position within the complex system that is emerging? The greater the degree of reflexivity in the authorial function, the more the reader is made continuously aware of the reading endeavor. In fact, the novel that overtly manifests its design must enlist the assistance of the reader in completing the conceptual circuitry between the plan and its execution. For example, the correspondence between "the author," John Barth, and the Lady Amherst in *LETTERS* may be compelling as an example of delicate social negotiations and epistolary style (both dying arts in our time), but the installment in this exchange is fictively meaningless—merely "droll"—unless the reader establishes its place in the construction of the novel as a whole. The "meaning" of the exchange is no more than an expression of its form. So "the author," acknowledging his own fictiveness, writes to his character, "I am by temperament a fabricator, not a drawer-from-life. I know what I'm about, but shall be relieved to get home to wholesale invention" (193). Proceduralists share a penchant for "blatant artifice" in literature with other practitioners of metafiction. They are allied with the surfictionists, especially in the use of graphical typography and the manipulation of textual space that one finds in Raymond Federman's *Take It or Leave It* (1976) or Ronald Sukenick's *Long Talking Bad Conditions Blues* (1979). In this regard proceduralism is always antimimetic, a contrivance that disturbs a reader's willingness to become absorbed in the regime of representation.

Some constraints are more covertly self-administered by the author. The first installment of Sorrentino's trilogy, *Odd Number*, borrows its title from Flann O'Brien's comic novel, *At Swim-Two-Birds* (1939): "Evil is even, truth is an odd number and death is a full stop" (314).[7] The reader should be thus forewarned that some use of numerology will figure in the novel. An unnamed interlocutor asks three sets of questions, deposing three apparently different respondents regarding the attendees of a party and, because conflicting evidence is given, the possible demise of Sheila Henry. If truth is revealed as an odd number, says O'Brien's Pooka, then the reader should be counting, as each section consists of thirty-three questions. There is never any suggestion that the interlocutor is aware that the three interrogations consist of an equal number of questions. But for the reasonably attentive reader, the hidden order emerges in the reading. It is a simple rule, and not overly difficult to confirm. Yet as one proceeds through the disorderly and conflicting responses that would seem to strike through almost every "fact" regarding the partygoers, there is an uncanny sense of an underlying pattern. Were the novel like a periodic attractor, a pendulum, passing over the same points repeatedly, the facts of person, place, or causality could be confirmed. The covert constraint of thirty-three questions, however, functions like a strange attractor in a chaotic system. No orbit through the system recurs on exactly the same trajectory, yet from the many trajectories a definable pattern in phase space emerges. No single fact can be determined or predicted, yet the design of the system reveals itself in time.

Elegant constructions of order and restraint are intrinsically appealing to the proceduralists, some of whom choose to work in plain sight of the reader, and some of whom prefer to operate from behind the arras. Nearly as common is the novel in which the protagonist attempts to impose a maniacal order on the chaotic conditions of his life. Samuel Beckett's *Watt* (1953) serves as a template for novels in which an eponymous character searches for an ordering principle that will lend some meaning to his life, and failing that, at least hold off insanity and institutionalization. Throughout the novel the hyperrationalist methods that Watt adopts to negotiate between his physical condition and his understanding of the world meet with unsatisfactory results. When the penurious Watt obtains mismatched footwear, the boot, size twelve, and the shoe, size ten, create considerable discomfort for his feet—inevitably, a size eleven. Watt tries to employ the sound constructions of logic to the incorrigible irregularities of life:

> By wearing, on the foot that was too small, not one sock of his pair of socks, but both, and on the foot that was too large, not the other, but none, Watt strove in vain to correct this asymmetry. But logic was on his side, and he remained faithful, when involved in a journey of any length, to this distribution of his socks, in preference to the other three. (219)

As in so many other situations in a novel that travesties the rationalist enterprise, Watt arrives at the correct solution to the number of ways in which his socks can be distributed on his feet (four), but this analysis fails to remedy the difficulties posed by a world in which irregular conditions are the norm. Watt's failings are ultimately surrogate for Beckett's own acknowledgment of the ineffectuality of language, and literary form, to bring an inchoate world into order and meaning.

Following in this tradition of futility, B. S. Johnson's hero in *Christie Malry's Own Double Entry* (1973) arrives at his "Great Idea," the application of double-entry bookkeeping to perceived slights against him (debits) and any benefits in his favor (credits) that the world might bring. The novel is punctuated with several Reckonings in which the dual columns of aggravation and recompense are tallied. Christie's mother remonstrates that the belief in a final reckoning, in which all injustices are evened out by the Chief Accountant—except by accident—is unfounded. She advises her son from her deathbed:

> "It seems that enough accidents happen for it to be a hope or even an expectation for most of us, the day of reckoning. But we shall die untidily, when we did not properly expect it, in a mess, most things

> unresolved, unreckoned, reflecting that it is all chaos. Even if we understand that all is chaos, the understanding itself represents a denial of chaos, and must be therefore an illusion." (30)

Christie's efforts to balance the books on his own life take the form of anarchist acts of violence against his employers, the state, and eventually the populace at large. Mother Malry's skepticism regarding the imposition of arbitrary order on the underlying chaos of reality is repeated nearly verbatim by Johnson in the introduction to his short fiction, *Aren't You Rather Young to Be Writing Your Memoirs?*, also published in 1973. (Perhaps determined not to die "when we did not properly expect it," Johnson committed suicide later that year.) No longer able to lend credence to a nineteenth-century reality of pattern and eternity, he suggests that "what characterises our reality is the probability that chaos is the most likely explanation; while at the same time recognising that even to seek an explanation represents a denial of chaos." In turn, he quotes Beckett as the only reliable authority, who calls for a new form in art which "will be of such a type that it admits the chaos, and does not try to say that the chaos is really something else. . . . To find a form that accommodates the mess, that is the task of the artist now" ("Introduction" 6–7).

Finding the forms that will accommodate chaos becomes the principal challenge to the postmodern novelist. As I will suggest in Chapter 6, Robert Coover's mild-mannered accountant, J. Henry Waugh, whose invention of a tabletop baseball game increasingly takes over his life in *The Universal Baseball Association* (1968), has much in common with Christie Malry, especially in his obsession with record-keeping, balancing the books, and evening the score for his diminished condition in life. The unnamed protagonist in Mathews's *The Journalist* likewise devises an elaborate system of classification for his daily activities—either verifiable or subjective; concerning others, or concerning only himself; and so on. Forgoing his prescribed antidepressants, he relies instead on the increasingly ramified categories of experience to organize his life. Yet the more invested these protagonists are in their perfectly ordered systems, the more susceptible they are to physical and psychological disorder and collapse. In this regard the characters who invent orderly systems for their worlds serve as surrogates to the postmodern novelists faced with the challenge of inventing literary forms that are not an imposition on but an accommodation of the disorderly processes of life.

Before turning to a discussion of three works of American proceduralism, let me characterize the necessary components of a procedural text. The success of these texts *as fiction* is inseparable from their success as literary devices. One cannot distinguish between a novel that satisfies as fiction but disappoints with gimmicky devices or a poorly executed construction; nor can one marvel at the elaborate structure that the text reveals but remain indifferent to its expression. All procedural texts share the following traits: invention; constraint; generation; and synergy. Members of the Paris Oulipo recognized both a historical cataloging of formal devices, an "analytical Oulipo," and experiment with new methods, a "synthetic Oulipo."[8] Postmodern American writers emphasize the inventiveness of their fiction. The design of the novel as a whole, and any constraints that it may employ, must be new; or, in the parlance of showmanship, never-before-attempted-by-human-hand. The design of the novel must also be arbitrary, a manifestation of literary artifice and unmotivated by any resemblance to theme, character, or the author's biography. (A design so motivated would be little more than a motif.) And as a corollary to its uniqueness, the design of a procedural fiction is for use one time only. Although it's conceivable that a design that produces an effective fiction in the hands of one novelist might produce a second, or might produce a novel of equivalent merit in the hands of another, in practice this replication is unheard of, because the form of invention is so wholly integrated into the practice of composition: the making and the writing are one.

Marcel Bénabou, in his essay "Rule and Constraint," draws a distinction between rule and constraint. Rules are associated with traditional literary production. A genre such as the historical romance, or a poetic form such as the sonnet, is governed by rules. They are the naturalized signs of the writer's mastery of technique or "craft." The author may follow these norms of composition unconsciously, and the reader may only be aware of their presence as they contribute to the pleasurable experience of reading. As I have suggested, rules are iterative; any writer may repeat them in practice. The constraint always presents the writer with a degree of difficulty that is exceptional. It is a restriction that presents a challenge, and requires exceptional skill to resolve. Bénabou suggests that "to the extent that constraint goes beyond rules which seem natural only to those people who have barely questioned language, it forces the system out of its routine functioning, thereby compelling it to reveal its hidden resources" (41). Unlike rules, constraints are always deliberately set by the author. They are, as Motte tells us, "*consciously* preelaborated and *voluntarily* imposed systems of artifice" (*Oulipo* 11). As intimidating as this procedure may sound, Bernardo Schiavetta insists that the literary constraint should not evoke the negative connotations of "obligation, hindrance, and lack of freedom." On the contrary, the constraint is freely adopted in a playful or ludic spirit. Here we encounter the paradox of the constraint, that what appears to be an unnecessary and deliberate restriction on the writer's practice can actually serve to liberate the writer from conventional means of expression and the tyranny of having to invoke some personal font of inspiration.[9] Sorrentino, for instance, regards these "formal structures as permissive of and conducive to compositional freedom" ("An Artist" 95). The writer may think of the constraint as a form of untracking or displacement from normative expression. Mathews says, "There is no value inherent in the product of a constrictive form, except one: being unable to say what you normally would, you must say what you normally wouldn't" ("Vanishing Point" 312). Procedural texts rely upon the constraint to generate the content of the work, propelling the writer toward new expression. Mathews, citing Jacques Roubaud, assigns the constrictive form with three responsibilities: "it defines the way a text is to be written; it supplies the mechanism that enables the text to proliferate; ultimately, it gives that text its meaning" ("Vanishing Point" 312). Without the capacity to generate text, it's fair to say, the constraint would be little more than a mechanical exercise.

I will eventually turn to more detailed descriptions of the constraints employed by Sorrentino, Mathews, and Barth in their fiction. But first I need to mention the importance of synergy in procedural works that function as complex systems of language. Each of these novelists has invented structures that can generate the text; they have voluntarily committed themselves to predetermined systems of artifice. Yet there is no manufacturer's warranty on literary proceduralism. The author may be confident that the adoption of a literary device will produce results, but the quality of those results may not be wholly foreseeable. Both the pleasure and the risk of the procedural text resides in this uncertainty. The author might formulate a design that produces a text with little more than a one-to-one correspondence to the effort expended. Such systems tend toward an unpromising steady state. Or the author might invent a structure whose agents achieve a combined effect that is greater than the sum of their individual effects. The steady-state design may suffer from overly rigid apparatuses of authorial control. Synergy in literary design arises when the complex dynamical system initiated by the constraint has the capacity to exceed authorial control. Even though the author is responsible for the *a priori* establishment of the constraint, the unpredictable nature of the language system may result in a creative autonomy, a generative text that far exceeds the enumeration of its preordained structure. In the synergy of design the text may write itself.

It may take a moment to appreciate the manner in which restrictions on the writing process might be liberating, and how constraints applied to composition might be productive of further writing. At the base of these counterintuitive claims is the proposition that simple rules lead to complex results in dynamical systems. If we accept that language—though artificial—is a dy-

Figure 3. The Mandelbrot Set (from Gleick, *Chaos*)

namical system, then predetermined rules applied in practice will produce unpredictable results. Investigations in chaotics show how this may be the case. In the field of deterministic chaos, James Gleick points out that the "Mandelbrot set is the most complex object in mathematics," and yet a complete description of this fractal can be transmitted by computer in only a few dozen characters of code (221). The infinite complexity of Benoit Mandelbrot's fractal is contained with a finite boundary. Resembling a hirsute Buddha or a tangle of sea horse tails under magnification, the Mandelbrot set is defined by self-similarity at different scales, though it never precisely repeats itself (Figure 3). Fractals are a testament to Mandelbrot's conviction that simplicity breeds complexity (222). Simple rules (which nevertheless require computers to carry out through unlimited iterations) generate these complex shapes. "For the Mandelbrot set," says Gleick, "the calculation was simple, because the process itself was so simple: the iteration in the complex plane of the mapping $z \rightarrow z^2 + c$. Take a number, multiply it by itself, and add the original number" (227). The result then becomes the starting number of the next iteration of the looping calculation.[10] "If the total runs away to infinity, then the point is not on the Mandelbrot set. If the total remains finite (it could be trapped in some repeating loop, or it could wander chaotically), then the point is in the Mandelbrot set" (223). Procedural fictions resemble fractals in that they are complex systems that rely upon feedback mechanisms. Output is always recycled into the system as input, changing the system for each subsequent iteration. Thus it is possible for mathematician or novelist to preestablish the rules that govern the system, but not be able to predict or control the patterns that result when the system is run. Most literary constraints, like Mandelbrot's calculation, remain fairly simple. The contents page of Barth's *The Last Voyage of Somebody the Sailor* describes a feedback loop: within the bounded limits of its framing narrative, the seven Voyages of the contemporary first-person narrator, Simon William Behler, are interspersed with first seven Interludes at the house of Sindbad the Sailor in medieval Baghdad, then six, declining by one until there are but one tale and one teller left. The lipogrammatic rule of Perec's *La Disparition* is a simple prohibition in the text of one, albeit the most commonly used, letter in French and English. Sorrentino imposes the constraint of an odd number of questions, and the order in which they are asked, on *Odd Number*. The enactment of simple rules in a nonlinear system effectively creates complex structures in time.

This relationship of simplicity and complexity is not limited to deterministic chaos. The biologist and complexity theorist Stuart Kauffman describes his research in autocatalytic sets in comparable terms. Darwin's theory of random mutation in the genotype is inadequate, Kauffman argues, to explain the evolution of the prolific order and variety that we find in the natural world. Rather than rely on "historical contingency" or "design by elimination" (7), Kauffman proposes a force of immanent design, which he calls "order for free," that stands in counterpoint to the law of thermodynamic entropy. Even with millennia at its disposal to sort through billions of adaptive variations, nature would be unlikely to arrive at complex organisms purely by accident. Kauffman bases his theory of coevolution, a "self-organization that arises naturally" (71), on autocatalytic systems that sustain and reproduce themselves without a genome. In *At Home in the Universe*, he speculates:

> If we are to believe that life began when molecules spontaneously joined to form autocatalytic metabolisms, we will have to find a source of molecular order, a source of the fundamental internal homeostasis that buffers cells against perturbations, a compromise that would allow the protocell networks to undergo slight fluctuations without collapsing. How, without a genome, would such order arise? It must somehow emerge from the collective dynamics of the network, the coordinated behavior of the coupled molecules. It must be another case of order for

free. . . . Astonishingly simple rules, or constraints, suffice to ensure that unexpected and profound dynamical order emerges spontaneously. (74)

Kauffman makes his case through studies of Boolean networks in which enzymes, figuratively represented as light bulbs, are turned on or off depending on the input from the neighbors to which they are "wired" at any given moment. Boolean functions such as AND and OR instruct an enzyme whether to be either active or inactive in a state cycle. Kauffman contends that researchers may control whether networks are "in an ordered regime, a chaotic regime, or a phase transition regime between these, 'on the edge of chaos,'" by adjusting the connectivity (the number of inputs) of the network and by altering the control rules (the Boolean functions) in the system (80–81). Similarly, the writer who implements constraints affects the collective dynamics of the fiction by determining the degree of restriction or difficulty and inventing control rules to direct the work into an orderly and/or disorderly regime.

Complexly Unrealistic: Gilbert Sorrentino's *Pack of Lies*

Gilbert Sorrentino has said that almost all of his thirteen works of fiction are written "under the influence of some sort of preconceived constraint or set of rules" ("An Artist" 95). The influence of proceduralism is especially apparent in the novels written after *Mulligan Stew* (1979), which has been described as a deconstruction, parody, or metatrope of the conventions of modernist fiction. Though his writing continues to acknowledge the influence of (and steal characters from) the novels of James Joyce, Samuel Beckett, and Flann O'Brien, Sorrentino's adoption of conspicuous design in his novels since 1980 indicates a postmodern turn. Perhaps the most widely known of his novels after *Mulligan Stew, Aberration of Starlight* (1980) presents a prismatic distortion into four parts of the events at a boarding house in Budd Lake, New Jersey over a thirty-six hour period in the summer of 1939. The divergence in point of view regarding an evening date gone awry between divorcée Marie Recco and the philandering salesman Tom Thebus, her frustrated father John McGrath, and disappointed son Billy Recco, is not estranged from the unreliable subjectivity of much modernist fiction. But the four narratives are further divided into ten sections, each in a different discourse, including a smug letter to friends at home, a utopian fantasy, catechistic interrogation (borrowed from the Ithaca chapter of *Ulysses*), and a random collage of the character's thoughts. The formal and stylistic parallelism among the ten sections indicates a convergence of vision that is finally the artist's. Structure, rather than the mundane narrative, governs the novel. In more recent works, even the pretense of narrative is abandoned. *Under the Shadow* (1991), Sorrentino tells us, "has a structure based upon the drawings done for Raymond Roussel's *Nouvelles Impressions D'Afrique* by H. A. Zo, drawings that, incidentally, have nothing to do with the text, but which, oddly enough, make a text of their own, a fragmented and discontinuous one, but a text nonetheless" ("An Artist" 95). The more arbitrary the structure, unmotivated by considerations of plot, the more conducive that constraint is to the generation of its own text.

In his essay, "Genetic Coding," Sorrentino acknowledges his joint Sicilian and Irish heritage. He appreciates that the art produced by these two peoples shares a "relentless investigation into the possibilities of form, a retreat from nature, a dearth of content" (264) that stems in part from the poverty of means they have endured. He is most influenced by those writers who possess these traits: "an obsessive concern with formal structure, a dislike of the replication of experience, a love of digression and embroidery, a great pleasure in false or ambiguous information, a desire to invent problems that only the invention of new forms can solve, and a joy in making mountains out of molehills" (265). These are clearly the codes by which he constructs his major work of the 1980s, the trilogy *Pack of Lies*, comprised of *Odd Number* (1985), *Rose Theatre* (1987) and *Misterioso* (1989).

Any work described as a trilogy—especially one categorized for the Library of Congress as "Detective and mystery stories, American"—raises certain expectations of continuity in narrative, the development of plot points through suspensions and resolutions, and the extended relationship of characters. Sorrentino's *Pack of Lies* makes mocking reference to these readerly expectations. The names of characters reprise in each of the novels, but in violation of literary convention, the characters are assigned different attributes. Each is a shifting signifier in a complex language game. One will have to find some other rationale than linear (narrative) sequence for grouping these three novels together, in the order in which they appear. Any suggestion that these novels follow from one another, that their contents progress in some fashion, is utterly defeated. And yet a trilogy in form it is. The author satirizes the literary critic's obsessions in *Odd Number*, when the second informant identifies Ann T. Redding as the author of "a book of criticism, literary essays" on the Janus theme, "one of those standard academic essays with a system, you know? Everything fits into the system or you don't mention it" (104). At the risk of seeming overly enamored with my own ideas of system, I want to propose that Sorrentino's structuring of the trilogy can be better understood with respect to chaotics. If Sorrentino, as he says, desires to "invent problems that only the invention of new forms can solve" in his novels, then *Pack of Lies* presents the problem of the nonlinear trilogy. Again, we would normally expect the parts of a trilogy to be connected, from A to B, and from B to C, and their orderly progression of linear causality to offer a satisfying sense of the whole. But the form of *Pack of Lies* emulates the orderly disorder of the fractal. Without threatening the reader of *Rose Theatre* any further with "a treatise on fractals" (154), consider the form of the Mandelbrot set, which displays self-similarity at different scales. Illustrations of the fractal created by Heinz-Otto Peitgen and Peter H. Richter (see Gleick, following 114) show that the hirsute Buddha that fancifully describes the figure of the whole set appears at various levels of magnification to about one million in each direction. Without coloration of the frames, and insets to guide one's sense of location, it is very difficult to distinguish the macroscopic view of the whole set from the microscopic views of the finer and more complex details with their Eiffel Tower spikes and swirling sea horse tails. In fractal geometry the form of the whole is implicate in the parts; and each part would seem to contain not one but innumerable wholes. Conventional distinctions between interior and exterior structures collapse, turn inside out and outside in.

There are myriad signs in *Pack of Lies* that the nonlinear, fractal geometry of the Mandelbrot set provides a more appropriate model for the trilogy than the linear sequence. The tripartite division of the whole is already contained within that of *Odd Number,* whose sections are in turn comprised of thirty-three questions apiece. The novel is replete with references to multiples of three: "three girls" (140), "thirty-three stories" all plagiarized (137), the title 99 (135), and so forth. *Rose Theatre* begins with a "Chayne of dragons" and ties several looping knots through the characters and contents of what precedes and follows it. *Misterioso* reprises in alphabetical order the names of characters, titles, and other substantive nouns from the previous two volumes, at once consuming and being consumed by its predecessors. Not unexpectedly, the three novels contain ironic and reflexive references to one another and to the trilogy, each seeking to contain the other. The interrogator in part three of *Odd Number* asks, "What titles might best describe the study I will undoubtedly write on these people and their relationships?" The informant's list of thirty-three titles includes "Doubles Cross, Odd Numerals" and "Complicated Webs" (144). The favorite book of Harlan Pungoe, a fraud, forger, and extortionist, is naturally *A Pack of Lies* (144), which he has committed to memory. *Misterioso* appears as "the new Broadway hit" (444); "*Odd Number* or *Odd Numerals*" appears as "an unreadable coterie novel" (463); *A Pack of Lies* returns as an anthology of "the new, unfrightened American realism" ready to conquer metafiction (470); and the play *Odd Number* is performed "in the old Rose Theatre" on Washington Square (501); all occur in, of course, alphabetical order in *Misterioso.* These reflexive references to the novels of the trilogy, and to the trilogy itself, in *Pack of Lies* amount to more than rampant intertextuality. The reader experiences much the same sense of vertigo that one has in viewing fractals: without fixed frames of reference—an inset picture or indication of the power of magnification—it becomes impossible for the viewer to determine the scale of the structure. Are we viewing an image of the whole contained within some infinitesimally small filament, or is the whole structure revealed to us in all its implicate form? In *Odd Number* the first informant provides a brief synopsis of *La Bouche métallique* or *Mouth of Steel* (see Sorrentino's *Steelwork* [1970]) as "an ingenious novel of worlds within worlds" (33).[11] Brian McHale contends that postmodern fiction expresses an ontological dominant (*Postmodernist Fiction* 10), an indeterminacy of being that compels the reader to inquire continually, "which world are we in now?" Sorrentino's trilogy eschews temporal sequence, linear causality, and ontological surety for a fractal form of self-similarity at different scales. *Pack of Lies* is an ingenious world of novels within novels.

The basic constraint suggested by the title of *Odd Number,* a division of the book into three sections comprised of thirty-three questions, generates for the author a sense of "the wholeness of the work itself" ("An Artist" 95). The preestablished number of ninety-nine questions serves as both motivation toward and a sign of the novel's completion. Sorrentino is at liberty to count other things—such as the list of thirty-three titles for the investigation (144)—and he liberally exercises his freedom under the constraint. The decision to count questions instead of pages, paragraphs, or words is an expression of authorial choice within the voluntarily imposed rules of the system. That choice, in turn, serves to direct the novel toward a particular discourse. What literary genre could be more conducive to a repeated series of questions than the detective or mystery story? The unidentified interrogator, it should go without saying, is not to be confused with the author. The impersonal register of the questions throughout the three sections of the novel invokes an official deposition by a district attorney, perhaps, seeking an indictment for manslaughter in the death of Sheila Henry: "What were you doing in the car?" (10) or "Why were you invited to this party?" (11). At the same time the reader understands these neutral questions to be the procedural device of an author intent on allowing the text to perform. The author who is inclined to invention will not be, as Barth says, "a drawer-from-life."

The three sets of thirty-three questions in *Odd Number,* like the trilogy itself, do not proceed in a manner that could be described as a logical sequence. One can speculate that they have been randomly drawn from some source known only to the author. The odd number evokes an instability or irresolution that seems calculated to undermine the conventions of detective fiction. Such novels may be serialized, but their cases are never left unresolved. It will be near impossible to ascertain the facts regarding "The Party," its occasion, attendees, and even its ontological level (is it a film, a script in a manila folder, or did it actually "occur"?), from the answers provided by the three informants. The first informant, who has access to extensive notes and data regarding "The Party" or *The Party* that he's collected in numerous folders, reacts defensively—as if to avoid any direct implication in the events he's describing—while responding to the interrogator in a syntactically disjointed fashion. He remarks on the illogical sequence of the questions, protesting that he could not have provided any earlier information about the incident (19). He objects to the interrogator's suggestion that he was a passenger in the car that contained Lou and Sheila Henry and Guy Lewis (10), and denies having been invited to any party (11). This instability in the questioning, noted by the first informant, is made more pronounced by the repetition of the questions in reverse order in the second part of the novel. The repeated questions make it clear that the interrogator has turned to a different informant, one who is now loquacious and eager to embellish his responses with irrelevant digressions and personal associations that have no bearing on the case, whatever it may be (Mackey 19). The answers to the questions are naturally quite different, with one exception: neither respondent has any idea what the phrase "metallic constructions" might mean (*Pack of Lies* 18, 106). The same question is later put to the third informant (the only instance of such repetition), who is perfunctory and unimpassioned when he answers: "*Metallic Constructions* is an inaccurate, or perhaps more fairly, an unsatisfactory translation of a technical work on engineering, *Les Constructions métalliques*, by Gaspard Monge, the inventor of descriptive geometry" (140). A clue to the importance of construction in Sorrentino's fiction is hidden in this insistence on the mysterious phrase. The interrogator next asks the third informant, "Can you tell me why I was directed to ask my fifth informant the same questions, in reverse order, that I asked my third informant?" (140). To which he replies curtly, "No." While the reader ponders the multiplying number of depositions, she might well ask the same question of herself: what is the purpose of the reversal of the order of questions? Like the first informant, the reader is already doubtful regarding the forward progress of the "narrative" in *Odd Number*. At the point at which the reader encounters the final question in the first deposition, "Why is it that so much of what you've told me also happens, more or less, in *Isolate Flecks*?" (58), and its repetition as the first question in the second deposition, the novel twists on itself and reverses direction. The astute reader will invariably refer to the answers presented in the first section—in effect, reading "backwards" in the novel—even as she proceeds in reading ahead in the second section. This reversal, or folding of the novel in upon itself becomes reflexively the subject of the two informants' responses. The first splutters, "if you think that what all this   all this is   is just some   from some fucking novel   why don't you read the novel and get all the answers right there?" (58). The second identifies *Isolate Flecks* as a pathetic *roman à clef* by Leo Kaufman involving the same personnel who appear in/at "The Party."[12] The twist that sends the reader moving simultaneously forwards and backwards can be found at the center of another mathematical figure of infinity within a confined space, the Möbius strip.

In an objection to the method of interrogation, the first informant complains:

> there's nothing   absolutely nothing   nothing that I've told you that hasn't been made into   made into some kind of a   what's that thing you do with a strip of paper where you twist it and then   you put the ends together?   so that you can't tell where   you can't tell one side from another?   what do they call that?   that's how you twist things   everything I say   you twist (49)

The Möbius strip is formed by rotating one end of a rectangular strip 180° and attaching it to the other end. *Odd Number* concludes with a continuous loop of text that would seem to describe the novel before us: "On his desk there is a manuscript, a typescript, to be precise, of a little more than a hundred and fifty pages. . . . Next to the manuscript is a single sheet of white paper on which there is typed a paragraph that reads:" (146). The text repeats itself up to the colon, implying repetition *ad infinitum*. But the text also forms a figure of infinite regress: an entire manuscript is described in a paragraph on a single sheet of paper; that paragraph in turn describes a manuscript, a typescript actually, that is described in a paragraph on a single sheet of paper; and so forth. *Misterioso,* the third novel in the trilogy, concludes with the sentences, "Twilight is approaching, yet another twilight. Soon it will again grow dark" (564), an apparent response to the initial question of *Odd Number,* "Was it still twilight, or had it already grown dark?" (9). The text of the trilogy begins again, one end attached to the other in an infinite textual loop like a Möbius strip.[13] Louis Mackey confirms that the Möbius strip is "a figure of the interminability and indeterminacy of representation. It never ends, and it has no respect for the law of contradiction" (23). But it is important not to forget that twist. The Möbius strip is a figural oddity because it is a continuous one-sided surface. Not only is it interminable, but as one proceeds around the strip, inside becomes outside and *vice versa*. The trilogy contains the three novels; but each novel in turn contains the trilogy. The stack of typescript includes the final paragraphs of *Odd Number;* the final paragraphs describe the entire typescript. *Pack of Lies* is very like a fractal, an infinite distance confined within a limited space, a form of self-similarity at different scales whose interior features are indistinguishable from its exterior features.

The third informant in *Odd Number* addresses the importance of a catalogue to the investigation with this thesis:

> If the catalogue, or any catalogue or list, is understood to be a system, its entropy is the measure of the unavailability of its energy for conversion to useful work.
>
> The ideal catalogue tends toward maximum entropy.
>
> Stick it in your ear. (144)

Rose Theatre, the middle of Sorrentino's trilogy, is constructed according to the principle of the catalogue. Although it sets out to remedy the testimonial distortion and correct the conflicting "data" of *Odd Number, Rose Theatre* shows itself to be a dissipative system. In a letter to his publisher at Dalkey Archive Press, Jack O'Brien, Sorrentino muses, "I have a feeling that a novel might be made of names only. I know one can be made of lists, and some day I'll write one. Maybe the next one, already in my head, based on an inventory made by Philip Henslowe of the Rose Theatre's props in 1598 in London, when he moved the company to a new location" (quoted in McPheron 66). True to his intentions, the fifteen chapter titles of *Rose Theatre* are borrowed from this list of props.[14] The fiction thus continuously refers to itself as a "found object," a catalogue (one of Sorrentino's favorite devices) of ultimately arbitrary, theatrical properties representing nothing more than the artifice of fiction.

In "Sittie of Rome," the final chapter of *Rose Theatre,* the narrator tells us that some unnamed individual is responsible for a "carefully designed plan, by means of which he is reasonably certain that you may discover various truths, or at the very least, certain insistent discrepancies among the data, however scattered, that one has been given to think of as factual" (282). This producer/director, possibly even the author, has brought together a tableau vivant of ten "middle-aged women" in a Roman café. Each chapter of *Rose Theatre* concerns the dispirited and sometimes abused women who make their initial appearance in *Odd Number*. Stylistically motivated by the English rose and Elizabethan trappings of the theater, the director presents his audience with a masque of women "he has been given to think of as actual" (282). They too are no more than properties of the theater, but that "is the gist of a scheme that he will proffer you" (283), the reader. Sheila Henry appears, in the chapter titled "Rocke," in a sepia-toned photograph seated on a large flat rock in a field. Yet rather than tell us what has become of Mrs. Henry after her unfortunate accident in *Odd Number*, the narrator considers the impossibility of the writer's task "to transform, or transcend life in order that it be made understandable, that's how art works, though he realizes that it is hard to understand life from a photograph, especially an imagined photograph" (232). As long as we understand the theatricality of all fiction, we can accept "the false news that Sheila had been accidentally run over by Lou" (275), suggested by the unreliable informant of *Odd Number*. We are further introduced in "Chyme of belles" (with pun intact) to Ellen and Anne, the two wives of Leo Kaufman, author of *Isolate Flecks*. Much of the "hastily scraped-together information" (204) appears in the form of a "promised inventory" (216) of traits, features, odd moments, miscellanies of worthless items, and useless facts. The more we know about these ten women, the less important the recollection of their "lives" becomes. The ideal catalogue tends toward maximum entropy in the system. The chapter, "Tree of gowlden apples," produces a family tree, a list of fancifully named children known to have sprung from the "ceaseless copulations of the group of husbands, wives, and lovers" (206), such genealogies being essential to the historical romance. About such lists the narrator observes:

> In conclusion, it may be seen that rigorous attention to the most pedestrian details of human relationships may yield surprising data if not any decent "yarns." Such data, while probably of little use to an understanding of the people involved in said relationships, may, however, allow us to draw certain conclusions about the truth behind the facade of social, public intercourse. Our paradigm is, we insist, despite the mockery we are all too used to, useful to a limited degree. Some may discern in it, perhaps, the presence of what Hans Dietrich Stöffel in his *De praestigiis amoris* (Brussels, 1884) defined as "[a] large mess (*perturbatio*) developing as if self-generated out of [a] small one." (207)

In this figure of a dynamical system it is possible to ascertain how the catalogue, list, or inventory (even one limited to the fifteen props from the Rose Theatre) may serve as a generative constraint. In systems far from equilibrium, large perturbations may arise from small; a butterfly's wing, a woman's glance, may produce a storm. This is the paradigm of Sorrentino's nonlinear narratives.

The source of the chapter titles for *Rose Theatre* in Henslowe's inventory is admittedly obscure, and might well not have been discovered were it not for Sorrentino's communication to his publisher. The constraint that governs *Misterioso,* which takes its title from the crepuscular song by the jazz composer Thelonious Monk, is not nearly so recondite. The alphabetical order into which the names of characters and places, titles, some substantive nouns, and other attributes appear to fall should be readily apparent to the attentive reader within a few pages at most. One is invited at the very start of the novel to speculate whether this arbitrary ordering will encompass the entire work as its principle of design, and whether the author will adhere to this scheme rigorously. Reading *Misterioso* entails the discovery of these questions of form; little else of narrative consequence exists to compel the reader. Sorrentino has employed an alphabetical schema in his book of prose poems, *Splendide-Hôtel* (1973), whose meditations are inspired by the letters that are displayed in dropped capitals at the beginning of each piece. One can also compare Walter Abish's *Alphabetical Africa* (1974), a travel narrative in which the first chapter, under the heading of "A," is composed only of words beginning with that letter. Each chapter adds another letter until, at the chapter titled "Z," the entire lexicon is made available. Then the constraint repeats itself in reverse order, subtracting letters until we return to "another Africa another alphabet" (152). The novel's constraint is both syntactic, affecting its composition, and structural, governing its form. Abish produces a strap-hinged encyclopedia of an imaginary dark continent. Sorrentino's *Misterioso,* however, has no chapter or section divisions, only blank slugs that separate text ranging from a single sentence to several paragraphs in length. The absence of narrative continuity or prominent division, in a novel that is as long as *Odd Number* and *Rose Theatre* combined, forebodes formlessness. I would argue that the alphabetical constraint in *Misterioso* is responsible for the orderly disorder of the text. The reader experiences the alphabetized materials of the text as blatant artifice, the imposition of an implausible ordering of persons, places, and things. In Roman Jakobson's structural linguistic terms, this would be an unusual imposition of the poetic axis of selection on the narrative axis of contiguity.[15] The novel begins with the discovery in an old A & P supermarket of a paperback copy of *Absalom! Absalom!* atop a binful of apples. Unlikely reading material for purveyors of the tabloid press, if indeed this is the work by that esteemed American novelist William Faulkner. What brings it there? Likewise, a few miles or pages away, a woman "in an Adirondack chair is of the opinion that she might as well be living in Akron, for, in the early afternoon, an Albanian woman of indeterminate age passes by" (291). What brings her there? These materials are brought into proximity by lexicographical accident, selected according to a principle that is foreign to the development of character, scene, or plot.

The simple rule of alphabetizing the first appearance of character and place names, titles, and other references compels unlikely juxtapositions throughout the novel. This constraint prevents an extended attention to a single protagonist, a locale, or protracted action around which a conventional narrative would accrete. In another letter to O'Brien, Sorrentino anticipates writing "a book that is a series of lists and catalogues—no narrative, no characters, no nothing but those words that 'tend toward maximum entropy'" (quoted in McPheron 68), referring to the theory of ideal catalogues expressed in *Odd Number.* Several narrative modules are periodically discernible in *Misterioso,* involving the patrons of the A & P, an alphabetical series of women from Karen Aileron to Karen Wyoming who are relentlessly chipper in "an intense sales-persuasion mode" (*Pack of Lies* 556), several episodes of *Buddy and His Boys on Mystery Mountain* which satirize the serial novels of moral instruction for youth, and the nightly entertainment available at the Bluebird Inn just outside Hackettstown, NJ on the road to Budd Lake, among others. The alphabetical constraint in *Misterioso* may not entirely dissipate the elements of fiction, but the text tends towards the entropic disorder that Sorrentino envisions.

The simple rule of an alphabetical order provides *Misterioso* with its complex design. In his reading of the "carnival of representation" in Sorrentino's novel, Louis Mackey observes an interesting anomaly in its form: there are not twenty-six sections, each one featuring names beginning with the appropriate letter of the alphabet, but twenty-five (63). The missing section—crossed out—should be devoted to X, Xavier to xylophone. Have the difficulties of maintaining the orderly procession of names finally taken their toll in the section with the fewest lexical resources in English? Mackey descries the missing X in the penultimate paragraph of the novel, mysteriously lurking among the Zs. Here we find two alphabetical series of paired attributes and names, one proceeding forward from the "actions performed by Deirdre Angelica" to "the zealotries championed by Marjorie Sciszek," the other working backward in counterpoint from "a glittering zipper opened by George Zaremias" to "the toy airplane dropped on Robert Armbrister" (*Pack of Lies* 563–64). "The two series intersect between M and N," Mackey observes, "and then go their separate ways. Dividing the Z entries, the two alphabetical rosters form a giant textual X that takes the place of the X omitted from the alphabet that organizes *Misterioso* as a whole" (64). While the reverse alphabet is perfectly ordered, the forward alphabet is slightly deformed by the auditory, but not lexical, initial zed of "Sciszek." The elegant but flawed design of this paragraph takes the form of one final list, recognizing that "many other curious and noteworthy things have happened today, involving people whose lives are, perhaps, as interesting or bland as those to which we have been constrained to pay heed, however superficially" (*Pack of Lies* 563). The author's constraint was, he tells Mackey in correspondence, "that all proper names previously used in *Odd Number* and *Rose Theatre* must, on their first occurrence in *Misterioso,* appear in alphabetical order" (Mackey 67). Sorrentino indicates that the deliberate anomalies in the alphabetical order of the text are examples of the Oulipian theory of the *clinamen*, or swerve. Georges Perec contends that a system too rigorously ordered will fail to be generative of any new order, remaining perpetually in an undifferentiated state. But a moment's turbulence, a slight imbalance in the system, permits the play that can produce new and unforeseen arrangements (Mackey 67).

Operating under its constraint *Misterioso* recapitulates the cast of characters from *Odd Number* and *Rose Theatre*, though not with the same *curriculum vitae* that we might have pieced together from those novels. Under B, Edward Beshary (now veteran of several Sorrentino novels) is the subject of "a long monograph in the *Enfer* division of La Bibliothéque Nationale concerning Beshary's perverse exploits while an assistant professor at Bennington" (*Pack of Lies* 302), amusing from we what know of Eddy Beshary's penchant for maladroit diction smattered with Brooklynese. And under H, the Henrys, Lou and Sheila, are reprised, he apparently homebound and given to wearing women's clothing, and she, sitting on a Coney Island beach, thinking of "skewed answers, obliquely given, to unasked questions" (377) until her "mind stops suddenly" (379). These passages, however, offer no culmination to the earlier novels. *Misterioso* enfolds the previous two works within its alphabetical ordering; but as in the folds of a strange attractor, no detail recurs in precisely the same place or with the same attributes. It is a complex mixing of recognizable elements into an entirely new pattern. The chaotic narrative and the elegant design of *Misterioso* are equally a product of the orderly disorder of the constrained text.

The Unclassifieds: Harry Mathews's *The Journalist*

Making his home in Paris, Harry Mathews has been closely associated during his career (his first novel, *The Conversions,* was published in 1962) with Georges Perec, Jacques Roubaud, Italo Calvino, and other members of the

Oulipo. He contributed an essay on Perec's remarkable life and fiction to *Grand Street* shortly after the French novelist's death from cancer in 1982. Mathews has translated several Oulipian writers from the French and co-edited an anthology of their writings, the *Oulipo Compendium*. As its only American member, Mathews brings to the Parisian literary association a more pragmatic approach to the difficult mathematical constraints that define Oulipian work. About his novel *Cigarettes* (1987), dedicated in memory of Perec, Mathews remarks, "It started as an attempt to solve a specific problem . . . of how to tell a story about a group of people belonging to the New York art and business world in a way that would allow the reader to make it up" ("A Conversation" 31). Mathews envisions himself as both a problem-*poser* and -*solver*, and his fiction enlists the Barthesian active reader in the resolution of the narrative conundrum that he presents. In a conversation with poet John Ashbery, he admits that, while it was the task of the Oulipo to "invent forms . . . that are very hard to use, very demanding, so that these will be available to other writers" as *potential* literature, in practice this invitation was rarely taken up. Many of these inventions remain "experiments in forms rather than in writing" ("John Ashbery" 43). Mathews organizes *Cigarettes* in an interlocking series of fourteen pairs of characters, between 1936 and 1963, among the social elite of Saratoga Springs with its thoroughbred horse track, and the aesthetes of Greenwich Village with its galleries and cafés. Edmund White comments that *Cigarettes*, though Mathews's "most realistic book," is also "the one most manipulated by combinatorial devices, such as those Mathews has described in an essay in Oulipo's *Atlas de littérature potentielle*" (78). The pairings of characters and other structural details in the novel are governed by algorithms like that described in "Mathews's Algorithm" for the *Atlas*.[16] Mathews prefers to withhold description of the specific constraints used in his fiction. But White observes that the "time signature" and "key" varies in each chapter of *Cigarettes*, e.g. "Allan and Owen, June-July 1963," according to the algorithm, while certain themes "repeat and evolve in 'free' (i.e., undetermined) fashion" (78) in the novel. The constrictive structure in *Cigarettes* is not *syntactic*, affecting the "material aspects of language (letters, words, syntax)," such as one finds in Perec's *A Void*, but *semantic*, affecting "what language talks about (subject, content, meaning)" ("Mathews's Algorithm" 136). The algorithm requires the use of a fixed set of "elements" in each chapter, and those elements are recombined in each succeeding chapter according to the "operation" described by the algorithm. The reader may only be able to reconstruct such an algorithm from the text with difficulty, but the evidence of its generative capacity in the narrative will be everywhere apparent, even if the constraint is not.

In contrast to the recondite algorithm of *Cigarettes*, the proceduralism of *The Journalist* makes itself plain within a few pages. Mathews describes his novel as "sort of half-Oulipian. I used Queneau's 'X takes Y for Z' to set up a field of relationships and then let the story work out of that. The narrative itself isn't Oulipian. . . . An Oulipian procedure can be anything from how you use a particular letter or a particular set of words to overall structures of long works, and there's a big difference between those extremes" ("An Interview" 43). *The Journalist* can be considered a companion to those novels, such as Alain Robbe-Grillet's *The Voyeur* (1958) and John Hawkes's *Travesty* (1976), in which an unnamed and unreliable narrator is entirely responsible for the account of the affairs that we are given. Like Robbe-Grillet's watch salesman keeping an incessant inventory of his wares, or Hawkes's driver calmly explaining his murder-suicide plot, the urbane diarist in Mathews's *The Journalist* presents himself under deep psychological strain and devises a plan for his own relief. Suffering from nervous exhaustion and depression, he omits his daily dose of lorazepam (a tranquilizer prescribed in the treatment of anxiety, nervous tension, and insomnia) in favor of keeping a journal of his activities on the recommendation of his psychologist. Beginning in the chronological order that one would expect of a diarist, his notes and reflections are soon cluttered with meticulously detailed lists of his expenditures, meals, meetings, and communications. The list, in the service of attesting to "facts," has its place in such masterworks of autobiographical writing as Henry David Thoreau's *Walden* or Benjamin Franklin's *Autobiography*. Intending his journal to be therapeutic, he is dismayed to find that he has inevitably "left things out" of his daily account. The list proves an inadequate solution because, he reflects, it "doesn't give me access to what it contains—its events are as dead as unrecorded ones. Its only benefit has been to show me how much I still leave out, how much more I want to get down, how I want to get *everything* down" (8). The impossibility of recording and organizing the seamless entirety of the quotidian quickly seizes upon the fragile attention of our narrator. Trying to rationalize even the ostensibly objective concerns of his life, he concedes that his account is "incomplete, false, and misleading. Itemizing the events of the day didn't save them for me. Objects and events, once I've *written* about them, emerge from the strangeness of belonging to systems outside my control" (9). The journal is a system of control that naturalizes its subjects only by wresting them from "the texture of life" (18).

Quickly dissatisfied with seriality, the journalist comes to a formal realization concerning chronological narrative: "How can I expect to include all I want from a day or part of a day I've just lived through if I meekly follow the line that leads from a beginning to an end? That line can only oversimplify.

It sticks to the obvious and reasonable, avoiding all that lies outside its 'inevitable' progress" (19–20). In the journalist's objections to chronology, one finds the basis for an author's resistance to linear narrative as the forcible arrangement of the continuum of experience into just such an inevitable progress of events. He decides to categorize his observations, "distinguishing between 'fact' and speculation, between what is external and verifiable and what is speculative. The distinction could supply the rudiments of an antichronological mechanism" (20). As he begins his increasingly ramified classifications of A. fact and B. speculation, the journalist becomes the surrogate author of a procedural novel. Devising his "antichronological mechanism" gradually becomes the primary subject of his deliberations and, to the detriment of his psychological health, his obsession. "Until I've learned to guarantee the accuracy of these pages," he declares, "my one concern must be perfecting my procedures" (156). Like Coover's J. Henry Waugh and Johnson's Christie Malry, Mathews's journalist embarks on an elaborate scheme intended to organize his life. The constraints are readily apparent to him because self-imposed. Naturally he admits to himself "a closeted vision, that of writing a novel—a novel about someone whose passion is keeping a journal" (209). Like the first-person narrator of John Barth's *The Floating Opera* (1956), Todd Andrews, who declares, "Good heavens, how does one write a novel! I mean, how can anybody stick to the story, if he's at all sensitive to the significances of things?" (2), yet proceeds apace to compose his multi-volume *Inquiry*, so Mathews's journalist overcomes his reservations about crossing from autobiography to fiction in producing the very novel we are reading.

All fables of intellectual inquiry feature a breakthrough moment, an exclamation of *eureka!*, and *The Journalist* provides such an instance at the close of Part I. Having already decided to "incerpt" or interleave his factual and speculative observations, those concerning others and those concerning only himself, the journalist announces, "While writing, I've realized something obvious. My categories (A1 A2 B1 B2) should be split into specific sections" (60). Once he begins the process of subdividing his categories to account for different types of observations, the journalist faces the unlimited prospect of finer distinctions that could be made to improve his classification system. Linnaeus's binomial system of nomenclature for the scientific classification of plants and animals has only to encompass the natural world; the journalist pits his system against the entirety of experience. He wonders "how thick this book will grow before the vase is broken" (60). Part II of the novel introduces marginal annotations that become increasingly ramified as the journalist refines his method. For example, when he records a dream (a subjective occurrence, involving only himself), the paragraph is appropriately indexed as

B II/a.1. A list of medical concerns (factual observations, concerning himself) is classified as A II/b.2*a*. Were the journalist actually the author of a novel (involving himself and others), its classification would fall under the rubric of metafiction. As the procedure for indexing his affairs grows more complex, the journalist's relationships with others and his treatment for depression are displaced by his concerns regarding the keeping of the journal itself. More and more of his time is devoted to bringing his "account of the day up to the minute" (70), until, as with Borges's Funes the Memorious, his waking hours are nearly consumed in the recording of his recollections. Charts depicting the branching structure of the "journal tree" are entered into the journal at various points in its development (84, 85–86, 112). The journalist confesses his doubts about the proposed system, "There's no denying that rearranging the incerpts is laborious, and each ramification will make it worse. Would indexing be more efficient? I must look into the possibility" (84). The reader of Part II of the novel sees the results of these procedural questions in the marginal annotations that accompany the text. Just as the journalist addresses himself with "maniacal obstinacy" (142) to his problem, so the reader of *The Journalist* is swept up in the fervor with which the affairs of a life might be ordered. One is encouraged to cross-check the indexing of the entries between the chart and the experience described, falling prey to the same obsessive compulsion that envelopes the journalist himself.

A cultured and well-educated individual, the journalist enjoys the operas of Mozart and Wagner, finds models for his writing in Plutarch's life of Pericles and a variety of European fiction, possesses epicurean taste in food and wines, and acquits himself (for a time) in business dealings and social engagements with the managerial class. Although the setting of the novel is never identified, one can infer from a weekend trip to a mountain lake district and the local fare served at the café ("A pile of deep-fried baby perch fresh from Lake John XXIII. Roast loin of pork with buttered new potatoes and a stew of the same yellow-stemmed brown-capped cantharellus I saw in the woods, here called September bugle" [46]), and the attendant characters of Swiss, Hungarian, and Czech nationality, that the journalist resides in a central European city such as Bern, Munich, or Innsbruck. He has an intrinsically orderly mind. In fact, he shares with the members of the Oulipo group an interest in mathematical problems. Working on his car, he considers the question, "how many ways can four wheels be positioned?" (35) in rotation. He swiftly determines that the "final number of permutations is thus twenty-four (4×6); arithmetically, the equivalent of $4 \times 3 \times 2$ ($\times 1$), or more generally, $n \times (n - 1) \times (n - 2) \times (n - 3)$ etc., where n = total number of units" (36), or n factorial. Later, in the kitchen, he examines the "possible combinations" in the preparation of an omelet (56). Despite his acuity in both culture and logistics, the journalist remains unaware of crises involving his family and friends. When a business associate, Stan, reveals that his young daughter has been hospitalized after a miscarriage, and that the father is "Somebody called Gert. She won't tell us his last name" (160), the journalist seems not to make the connection to his own college-aged son named Gert, though the reader surely does. His wife Daisy easily conceals her overnight hospital visit for a cancer scare, though he suspects that she's having an affair with their good friend Paul. He struggles with his solipsism, in the vernacular of his classification system (annotated as B II/b.3*a*):

> I was reflecting that in this task I've set myself I'm totally alone, and it cannot be otherwise. I have nothing but the vaguest help and support. How can I ever explain to anyone the high principle that governs my commitment—a commitment to the 'perfectibility' of my methods? It may now concern only myself, but ultimately it extends far beyond myself because it is the condition of the noblest of all acts, that of rescuing the precarious imprints of reality; and reality is the world's, not mine. (153–54)

The insoluble problem of the journalist's system for indexing lived experience is the presence of an always unclassifiable "other." Each subdivision of his categories makes finer distinctions, but each leaves behind an intransigent remainder of "other actions and events," "other matters," "other people" (84). He comes to realize that "it's not this or that category, it's the overall problem I can't master. The more I put in, the more I leave out" (110). Finally, the journalist must account for a residuum "other" in nearly every subdivision he makes (112). The problem is not unlike that of the scalar geometry of mapping. The smaller the scale of measurement, the more detailed will be the map. A scale of one-inch-equals-one-mile on the legend includes more features than a scale of one-inch-equals-ten-miles. But every mapping of the irregular shapes in the world only serves to suggest that there are still more features that would be revealed by a still smaller scale. The more one puts in, the more one leaves out. The journalist's cognitive mapping (to use Fredric Jameson's term in *Postmodernism*) of his world is destroyed by the impossibility of its every achieving a one-to-one correspondence with the enormous complexity of daily life. We live always imprisoned by the arbitrary limits we impose in organizing "a world" we can understand from the continuum of all possible worlds.

In what is both a cry of despair and another exclamation of *eureka!*, the journalist recognizes that "all the care I have brought to organizing this journal has been misspent; my laborious classifications have proved worthless; my efforts at competence are an illusion. Why? Because I have left out the chief activity of my life and the chief fact of my project: the keeping of this journal" (190–91). He is dissatisfied with the preliminary design of the journal because—in attending to the classification of fact and speculation—he fails to account for his writing of the journal. "In my elaborate system, what place has been set aside for the compilation of these pages? A niche in A II/b.3*b*" among the many "other" verifiable events concerning himself alone (191). The task of describing the system in its entirety can't be assigned to a subdivision of the system itself. Instead, he proposes keeping a "Journal of the Journal" (194) in which the act of recording every entry in the journal would have its corresponding account of the writing of that entry. The framing of the system would then require that he "make a frame for the frame" and lead swiftly to "a discouragingly infinite regression" (191). He acknowledges that to enter such a mirrored room would take him irrevocably away from friends and family—anything concerning others.

The journalist's manic obsession with making a perfectly ordered world leads to the system's dissipation and his own collapse. The medical men come for him, and he is admonished not to write or read further in his diary (230). But only the dissipation of the orderly system makes possible its reorganization at a metalevel. Were the flaws in its design not revealed in the straining toward perfectibility, the system would remain unchanged. Part III of the novel dispenses with the marginal indexing of the text in recognition that the Journal of the Journal lies beyond orderly classification. Although keeping a journal has not been salutary, the journalist won't resign himself to the sedate stability offered by his prescription lorazepam. He realizes that the "transient mundanities [the journal] records survive only through their presence here, while the page itself creates its own transcendent life" (191). Unapologetically admiring his work, he sees how "the complex design eloquently revealed the simple concept behind it" (211). Mathews's *The Journalist* is the perfect study of this principal of proceduralism.

Navigating the Periplum: John Barth's *Last Voyage*

In August of 1991—early in the Atlantic seaboard's hurricane season—John Barth ventured from his home along the Chesapeake Bay in Maryland to the inaugural Stuttgart Seminar in Intercultural Studies, to deliver a series of lectures in which (with some trepidation) he would suggest a sympathetic vibration between postmodernism, chaos theory, and the romantic arabesque. As a working professional novelist who reserves his Fridays for forays into aesthetic theorizing, Barth's account of the seminar and his revised lectures are included in his collection of essays, *Further Fridays* (1995) to which the reader is warmly directed. At that date Barth had just published *The Last Voyage of Somebody the Sailor*, which he describes as a "postmodern arabesque novel" (*Further Fridays* 278). I'll refrain from recapitulating Barth's discussion of the homologies between Jean-François Lyotard's theories in *The Postmodern Condition* and chaos theory as it has been described by N. Katherine Hayles and James Gleick, partly because I've already presented aspects of this relationship in my introduction. The third leg of the triskelion, the arabesque, reflects Barth's life-long fascination with the Arabic compendium of tales, *The Book of a Thousand and One Nights* (Kitab Alf Laylah Wa Laylah), and the three-thousand-year-old Arabo-Oriental tradition of design. Barth finds an "uncanny prefiguration of a meeting ground" for the arabesque, chaos theory, and postmodernism in Friedrich von Schlegel's application of the term *Arabeske* to the use of framing narratives in his theorizing about the genre of the novel (*Further Fridays* 284). Barth mentions his own recently completed novel only in passing (knowing that authors should never critique their own creations at the risk of incurring the Muse's wrath), though he describes his "working aesthetic" as "chaotic-arabesque Postmodernism" (289). The critic, however, is free to suggest that this proposition of a "mode locking" between aesthetic, scientific, and philosophical theories provides an excellent description of Barth's strategies in *The Last Voyage of Somebody the Sailor.*

Barth spies a convergence in Schlegel's definition of arabesque romance as a "structured, artful chaos" ("*ein gebildetes künstliches Chaos*") and James Gleick's characterization of chaos (in the field of nonlinear dynamics) as "an orderly disorder generated by simple processes" (*Further Fridays* 286). Note that both expressions flirt with a paradoxical relationship to describe phenomena that are not fixable in a single term. Gleick presents a branch of chaos theory, illustrated by fractals and strange attractors, that finds order hidden within chaotic systems. He emphasizes the deterministic rules that produce these complex phenomena. Schlegel's use of the adjective "structured" distinguishes chaos from mere disorder or entropy on the one hand, and from the orderly cosmos on the other; while "artful" differentiates between the structured chaos of art and that of nature. In accord with Barth's metafictive practice, Schlegel's expression "implies an artist behind the structure, an artful controller of the controlled indeterminacy" (286). Chaotics and the arabesque share the apparently paradoxical relationship of a structured disorder or controlled indeterminacy.

In his third lecture, Barth expatiates on the parallels between arabesque

Figure 4. Detail of Mandelbrot Set showing recursive symmetry across scales

design and Mandelbrot's fractal geometry. The arabesque style of design—including but not limited to the figuration found in "Persian" carpets, mosques, and the framed tales of *The Arabian Nights*—is characterized by "highly intricate patterns of geometric designs (floral or otherwise) involving structurally repeated, symmetrically developed lines, loops, concentric and interpenetrating curvilinear, triangular, rectilinear, and quincunxial structures" (312, quoting from Michael Craig Hillman, *Persian Carpets*). Among the more frequently repeated structures in carpet design is the interlineated quincunx, a figure with a central medallion in a framed field of four corner points. The medallion, with pendants extending toward but not quite touching the bordering fields, and the four corner elements whose intricate foliations are limited to their respective regions, resemble, to Barth's eye, Mandelbrot fractals (313). Several features of the arabesque and the fractal would seem to confirm this comparison. The foliations of arabesque design, because they are geometric abstractions of natural forms such as flowers and leaves, involve structurally repeated figures moving from the center to the border or vice versa. These repetitions strongly suggest the recursive symmetry across scales that distinguishes fractals (Figure 4). The comparison is not an idle one, because recursive symmetry (though not to infinity) can be found copiously in the natural world, from snowflakes to the curling leaves of ferns. In the "implicature of the *quincunx*" one discovers "repeated patterns within repeated borders to suggest infinity" (315, quoting G. R. Thompson, "Romantic Arabesque"). Barth studiously observes that fractal geometry "affords more dazzling instances of interplay between finite and infinite," such as the Koch snowflake in which a finite area is surrounded by an infinitely long border (315–16). Fractals are not small worlds of perfect order; rather they reveal the regular irregularity that is prevalent in the world. Similarly, deliberate flaws are said to be woven into Persian carpets in recognition of an imperfect world. Such flaws in design remind one of the *clinamen*, the error introduced into texts written under constraint that Perec argues contributes to its generative capacity. Barth finds that, because of the Islamic prohibition against reproducing natural forms (iconography), the "subject matter" of Persian carpets is ultimately the pattern itself. The arabesque reinforces the reflexivity that one associates with postmodern fiction. These fundamentals of arabesque design, shared by the orderly disorder of fractal geometry, provide Barth with a model for the design of his novel.

The Last Voyage of Somebody the Sailor reveals a quincunxial design—like the compass rose that graces old navigational charts—in its narrative structure. The central first-person narrator is Sīmon ("seaman") Bey el-Loor (Baylor, Somebody the Sailor), who tells the Seven Voyages of his life on seven consecutive nights as a guest in the house of Sindbad the Sailor in medieval Baghdad. To the NE lies aging and increasingly erratic-of-mind Sindbad, telling tales from his Tub of Last Resort about his previous Voyages, negotiating the funding for his Seventh and the dowry for his daughter's marriage. Opposite Sindbad, on the SW corner, we find Simon William Behler, near-exact contemporary of John Barth,[17] an award-winning New Journalist whose travels are recounted in his autobiographical reportage, who becomes "stranded out of time" on a sail from Sri Lanka (Serendib) to Basra. To the SE sits Yasmīn, the enchanting green-eyed daughter of Sindbad, whose travails at the hands of the pirate Sahīm al-Layl, and whose virginity (and the current disposition thereof) are endlessly discussed. On the NW corner reclines Behler's adolescent girlfriend, crazy Daisy Moore, who introduces him to the fragrant bower of sexuality, and all subsequent fulfillers of delight, wife Jane, the jasmine-scented watch-seller in Dronningen's Gade in the Virgin Islands, and later sailing partner Julia Moore. The design of the quincunx accommodates a series of twinnings that cross through the principal narrator, Sīmon, aligning the characters of Behler's tidewater town of East Dorset on the Eastern Shore of Maryland in the twentieth century and those who are borrowed from Sir Richard Burton's translation of *The Book of the Thousand Nights and a Night* (1885). Barth, who has a twin sister named Jill, provides for his surrogate, Behler, a fraternal twin who "didn't quite make it into this world" (*Last Voyage* 27). Behler loses his first partner on the sea journey into life, as he later loses Julia Moore off the modern-day *Zahir* in the Indian Ocean sailing from Sri Lanka,[18] and as Bey el-Loor finally loses Yasmīn on what could be called the return leg to Serendib. As the central medallion offers a prospect of infinity in arabesque design, so it is through the figure of Sīmon Bey el-Loor, stranded out of time, that the worlds of Arabian fable and contemporary America must pass, and in so doing confirm the timeless values that they share in common.

Though textile design is an important source of the arabesque, the collections of Arabic-language tales and poems such as *The Ocean of Story* and *The Thousand and One Nights* provide another model for the structure of Barth's *Last Voyage.* The tales themselves, written in a vernacular rather than a literary style, filled with fantastic voyages and pedagogical fables, were widely popularized in the West during the eighteenth and nineteenth century. Barth observes, however, that "for Western literary *theory* as opposed to popular literary fashion, the significant European appropriation from Arabo-Oriental literature and art history is not subject matter but design: 'arabesque' in the sense of elaborately and/or subtly *framed* design" (*Further Fridays* 318). Barth does appropriate the subject matter of Scheherazade's tales for *Last Voyage,* and can't resist the challenge to his masterful storytelling by supplementing narratives and embellishing characters in the classical repertoire. And yet he relies on an elaborately framed design in *Last Voyage* to weave the stories of coming-of-age in wartime America and the Arabic fables of fortune and survival into a complex whole. Barth quotes G. R. Thompson on this feature of the arabesque: "The framing of frames [as in 'Persian' carpets and Scheherazade's tales within tales within tales] suggests pure design" (318, Barth's brackets). For Barth the proceduralist, the aspiration to pure design defines the success of *Last Voyage,* because none of the tales recounted would be sufficient alone to sustain interest. The author's compulsion, and the reader's attention, lie in assessing the complicated arrangement into which the tales fall. Returning to Thompson, "By analogy to a sense of elaborately symmetrical yet open-ended design, 'arabesque' as a European literary term came to indicate deliberate inconsistencies in the handling of narrative frames and . . . intricate ironic interrelationships among many tales within a frame or a series of nested frames" (318, Barth's ellipsis). Apprehending and evaluating the nested frames in *Last Voyage* becomes both the necessary task and the pleasure of the text for the reader.

The outermost frame tale in *Last Voyage* places Simon William Behler in a hospital bed attended by a "familiar stranger," a green-eyed woman (4) who resembles—or may be—the impetuous Yasmīn of the Arabian tales. Or perhaps she is a visitation of his lost twin, "green-eyed BeeGee" (573) come to escort him into the next world. Like Odysseus washed up on the shores of Phaiákia, Behler begins to spin fantastic tales to his "Nausicaä-on-the-Chaptico" (571). This exterior border does nothing to resolve the verity of an outlandish story of a man stranded out of time. And nestled inside the frame tale situated in contemporary America we find a second Arabian frame, in which Scheherazade (having long ago finished her thousand-and-one tales) offers her last, to forestall death (The Destroyer of Delights) a little further. She then tells The Last Voyage of Somebody the Sailor; this invites the conclusion that what follows is no castaway's delirium but one more story to pass an Arabian night. Again, as Thompson informs Barth who informs us, "In the arabesque, the relationship between a framing narrative and one or more story strands severely strains or calls into question overt narrative illusion, which may be further undermined through involuted narrative conventions, complex digressions, disruptions or incongruities, and the blurring of levels of narrative reality" (*Further Fridays* 318), all of which effects are found abundantly in *Last Voyage*. In a narrative of voyages, with strandings both foreseen and unforeseen, "wandering back and forth between temporal and spatial settings" is Barth's *modus navigandi*. He tacks between the "homely Islamic realism" (*Last Voyage* 136) of giant rocs and jewel-strewn shorelines and the Tidewater fantasia of flying ships and magical timepieces that keep on ticking.

Digression and incongruity are essential aspects of the arabesque design of *Last Voyage*, but one does not want to convey the impression that Barth has charted a course for serendipity in this 573-page novel. The embedded frames of the arabesque aid the reader in establishing the organization of the novel. As we leave that safe harbor, we embark upon the Seven Voyages of Somebody the Sailor and the Interludes (beginning with seven, and declining by one until the last installment) set at the dinner table of Sindbad the Sailor. If we examine the narrative structure in the Last Voyage, recursive symmetry across scale levels appears as the story proceeds. Barth's Stuttgart seminars propose that the orderly disorder of chaotics is strongly suggested in arabesque design and postmodernism. Fractal shapes as well as "Persian" carpets provide the model for the novel's form. Voyage One and its seven following Interludes begin far apart geographically (in East Dorset, Maryland and Baghdad), temporally (on Behler's birthday on July 1, 1937 and at an undetermined moment in the twelfth to fourteenth century), and aesthetically (in realism and in fabulism). A disconcerting array of transitions, leaving much unexplained. Sīmon Bey el-Loor's strange tales of boyhood on the Chesapeake, of wartime tragedy and sexual awakening, are harshly criticized by Sindbad's dinner guests for their incredibility and their offense to Arabic custom. In Voyage Three, a family sailing trip through the Virgin Islands in 1972, Behler (now 42 years old) enters a Muslim-Indian owned watch shop in the free port of Charlotte Amalie. He purchases a Seiko with a rotating bezel from a young, jasmine-scented saleswoman, the doppelganger of Yasmīn. Adjusting the watch's bezel as he departs—or rubbing the magic "amulet" (367)—he finds himself lost in an Arabic souk, as if in a different century. This passage delights his audience—all the Voyages are recounted at Sindbad's table—and prepares the reader for the events of Voyage Five. Retracing Sindbad's return to Baghdad from the Far East with green-eyed Julia Moore, free-lance photographer and younger sister of Daisy, aboard a cutter re-christened the *Zahir* in 1980, Behler nearly drowns in the Indian Ocean in an attempt to save his mate who has been swept overboard. He awakens aboard the *Zahir*, a fiber-sewn boom piloted by Mustafā mu-Allim down the Persian Gulf on a mission to deliver Yasmīn to her betrothed. Behler's magical timepiece, his Seiko, becomes both a navigational aid and the talisman of any hope for return to the twentieth century. The narratives of Behler's Voyages and Sindbad's Interludes now begin to converge, in a fashion described by Yasmīn by tracing her initial on her belly: Y (366).

Simon Behler/Bey el-Loor is the "strange attractor" in the narrative of *Last Voyage*. One can think of these American and Arabian tales as two drops of dye, one red and one green, on the surface of a white liquid. As the liquid is vibrated, the two spots of color, initially separate, begin to stretch in a swirling motion around some central point, without entirely blending into one another. Edward Lorenz's strange attractor describes two spiral loops in phase space. Its lines are infinitely long, yet remain within a finite space. The loops never merge, and they never intersect (Gleick 140). All strange attractors are figures of infinite regression. Bey el-Loor testifies in the Fifth and Sixth Voyages to his experience on board the *Zahir:* the hijacking of the ship by the pirate Sahīm al-Layl, the ransoming of Yasmīn, and the revenge taken upon Sindbad by his foster son, Umar al-Yom (in his assumed identity as Sahīm). These accounts, and the Interludes that follow them, contain several disparate renditions of those events by the other principals—Yasmīn, Sindbad, Jaydā the duenna, and Sahīm/Umar. The tales converge in multiple iterations, but they never exactly repeat themselves. One instance of nearly infinite narrative regress occurs in Voyage Six. Bey el-Loor, deposited by the pirate vessel *Shaitan* in the Shatt al Arab near Basra, is reunited with Yasmīn. He thoughtfully tells her "the full story" so far. (Yasmīn is herself among the diners on Night Six, so she must once again be audience to this telling of the story.) Our hero tells of "Umar's version of 'Tub Night' [the details of whose clandestine meetings differ among all the accounts] and of her father's fifth voyage, wherein Sindbad himself had been the monstrous Old Man of the Sea [differing from Sindbad's telling in the interludes]; Umar/Sahīm's ongoing love and bafflement and rage; his mutilation and transformation" (515) and so forth, until Bey el-Loor returns in his narrative to Yasmīn's doorstep. Tales told within tales told within tales. Barth regards "the strange-attractor branch of chaology . . . as having more affinity with Postmodernism and the arabesque." Katherine Hayles's description of the Lorenz attractor that almost but never quite repeats itself "brings to mind arabesque carpet design, Friedrich von Schlegel's praise of self-similarity in Shakespeare, . . . perhaps even my own most recent 'arabesque'" (*Further Fridays* 335). As previously noted, Barth designs *The Last Voyage of Somebody the Sailor* so that, as the Interludes decline in number from seven to one and the tales of Sindbad and Somebody the Sailor converge without exact repetition, "the voyager is changed by the voyage, but the voyage is also changed by the voyager: dynamic feedback loops" (*Further Fridays* 334).

Barthes prefers his own coinage, "coaxial esemplasy," connecting the twentieth-century technology of fiber optics to nineteenth-century aesthetic theories of the imagination, to describe the importance of reciprocal influence in his novels. I have suggested that the relation of the author to the text in procedural fiction resembles that of a dynamic feedback loop. The appeal of design for Sorrentino, Mathews, and Barth lies not in the imposition of a deterministic order or the assertion of restrictive control over the text. Rather these authors are voyagers who are changed by the voyage. As the navigator assiduously charts the ship's position on his map, the actual course of the voyage is continually changing. Not incidentally, Norbert Wiener derived the term for his science of communication and control systems, cybernetics, from the Greek word, κυβερνήτης, for "steersman" (*Human Use* 15). The operator of the system relies on sensory feedback based on its actual performance rather than its expected performance. Modifying his behavior in response to the past performance of the system, the operator adjusts his directives to the system. Output becomes input in a continuous feedback loop. Wiener recognized that the operator had become part of the system. Dynamical systems situate the author as intrinsic to the text's design. Mathews contends that literary constraints prevent the author from employing familiar or normative expressions and so compel him to say what he normally wouldn't. He describes a textual system in which an expected performance must be adjusted to reflect the actual performance. Sorrentino's conviction that structure can generate content in his fiction relies upon the reciprocal influence between author and text. The author invents the structure of the work, but that structure compels his performance in ways that he had not anticipated. Procedural fictions that merely carry out a set of instructions are fine bits of engineering; those that induce a limitless exchange between author and text are true expressions of design.

5
Noise and Signal: Information Theory in Don DeLillo's *White Noise*

[In electronic music] sounds themselves will consist of unusual frequencies that bear no resemblance to the more familiar musical note and which, therefore, yank the listener away from the auditive world he has previously been accustomed to. Here, the field of meanings becomes denser, the message opens up to all sorts of possible solutions, and the amount of information increases enormously. But let us now try to take this imprecision—and this information—beyond its outermost limit, to complicate the coexistence of the sounds, to thicken the plot. If we do so, we will obtain "white noise," the undifferentiated sum of all frequencies—a noise which, logically speaking, should give us the greatest possible amount of information, but which in fact gives us none at all. Deprived of all indication, all direction, the listener's ear is no longer capable even of choosing; all it can do is remain passive and impotent in the face of the original chaos. For there is a limit beyond which wealth of information becomes mere noise.

—Umberto Eco, *The Open Work*

The noise is the background, the pebbly depth—granular, sandy, quasi-liquid, fluctuating—that underlies information; the shape is rare, like language. Language: the rose; noise: the pebble, the water with which one sprinkles the floor of the room.

Sprinkles, so that roses come, from noise to meaning.

—Michel Serres, "Literature and the Exact Sciences"

Don DeLillo graduated from Fordham University in the Bronx, New York in 1958 with a B.A. in Communication Arts. As an aspiring fiction writer, he may have chosen this field over the more traditional literary degree in English because it offered a pragmatic introduction to the broadcast media of television and radio or to print journalism, areas in which a young man with talent as a writer might conjure a living. In fact, DeLillo did publish a number of articles and essays as a free-lance journalist—an occupation he describes as exhausting, and from which his success as a novelist has offered salvation. He has never taught literature in the university, nor held any sort of academic position. One presumes that his program in the communication arts involved some component of information theory, which had been proposed by Claude Shannon only ten years earlier, since the conveyance of information would be the essential aim of communication. One approach to reading DeLillo's eighth novel, *White Noise* (1985), is as a catalogue of the malaise of the Information Age. As revenge on a master's thesis never written, DeLillo parodies his former classmates in Communication Arts who might have gone on to profess their knowledge of media culture in the academy. Both the academic and the domestic characters of *White Noise* suffer from the symptoms of "information sickness."[1] In earlier novels DeLillo describes this condition as "sensory overload" (*Great Jones Street* 252) or the "superabundance of technology" (*Running Dog* 93).[2] The Gladney family, who flee an Airborne Toxic Event in Part II of the novel, display no apparent signs of chemical or radiation exposure but suffer instead from the incessant and high-level dose of information that permeates their daily existence. DeLillo's fictional Geiger counter measures exposure not in *rads* but in *data*. With its origins in the Cold War period of backyard bomb shelters and school-time "duck-and-cover drills," the novel records the post-apocalyptic fallout of an enormously powerful information explosion that no one noticed had occurred.

Noise

The first domestic scene of Don DeLillo's *White Noise* dispels any sense that we are comfortably at home in what DeLillo has described as the "around-the-house-and-in-the-backyard" school of neorealist American fiction (Harris 26). Jack Gladney, the chairman of the department of Hitler studies at the College-on-the-Hill (one suspects this hyphenated institution is located in the conservative hinterlands of west-central Pennsylvania), attends to his family—the assembled spawn of several past marriages enjoyed by Gladney and his wife, the ample Babette. Although not wearing the academic robes that he dons on campus to lend himself an air of authority, he observes with some formality that "We entered a period of chaos and noise" (6). He is speaking portentously of lunch. In DeLillo's fiction the minute particulars of the contemporary American habitat—a kitchen littered with the disposable packaging from processed foods: "open cartons, crumpled tinfoil, shiny bags of potato chips, bowls of pasty substances covered with plastic wrap, flip-top rings and twist ties, individually wrapped slices of orange cheese" (7)—are interrogated for a message beyond their immediate function. Convenience, comestibility, unnaturally bright coloring. Gladney's precocious son Heinrich "studied the scene carefully" but departs without pronouncement. Presumably momentous events, however, such as the Airborne Toxic Event that descends upon the town of Blacksmith in Part II of the novel, remain ultimately uninterpretable. Confusion mounts as suspiciously reassuring but unreliable reports are published. For DeLillo, then, the "period of chaos and noise" that we have entered is more appropriately identified as postmodernism. The plotted curve of *White Noise* demonstrates the nonlinearity of cause and effect—or the indeterminism—of postmodernity. Whereas neorealist fiction abides by the deterministic presumption that the small niceties of living will retain an innocuous glow, and that the larger events will have profound repercussions in our lives, postmodern fiction engages the collapse of the mechanistic paradigm of linear causality. In open systems—including the chaos of the suburban kitchen—small causes may emerge through feedback or looping into effects of great importance; large causes may dissipate in their impact.[3] The catastrophic Airborne Toxic Event has no immediately apparent repercussions on the community of Blacksmith beyond a "level of experience to which we will gradually adjust, into which our uncertainty will eventually be absorbed" (324–25).

Likewise, the affection of postmodernity for kitsch is explained by the Gladney's lunch table.[4] An uncertain relationship develops between the sign and the object. The artificial additives in the "brightly colored food" (7) available for lunch designate its gustatory appeal. The sign of the orange Velveeta becomes far more compelling than the substance contained in the processed "cheese food" that no longer bears much relation to a dairy product. At the close of the Gladney family's postmodern lunch period, "The smoke alarm went off in the hallway upstairs, either to let us know the battery had just died or because the house was on fire. We finished our lunch in silence" (8). The messages and warnings of postmodern fiction contain a component of undecideability, careening between mental note and crisis alert.

This postmodern lunch period might also be designated as the Information Age. The so-called American Century has witnessed the shift from hot-and-heavy industry such as steel production and automobile manufacturing to the cool-and-light technology of software, automatic banking, and laser radial keratotomy. Winnie Richards, a neurochemist on the College faculty whom Gladney consults about his wife's bootleg psychopharmaceuticals, remarks that Americans "still lead the world in stimuli" (189). The Industrial

Age had certain advantages. Raw materials were a limited resource, and it was possible for an Andrew Carnegie to dominate the market in steel production and own most of Pittsburgh (cleverly disguised as Iron City in *White Noise*). A good piece of steel was readily identifiable. In the Information Age, however, both the producer and the consumer are awash in a superflux of information. It becomes increasingly difficult to distinguish between reliable information and meaningless noise. For Jack Gladney, both the riotous abundance of signals and the incessant background hum of noise present problems of interpretation.

In his book-length poem, *Garbage* (1993), A. R. Ammons suggests that "garbage has to be the poem of our time because / garbage is spiritual" (18). The garbage trucks tend to the enormous piles of trash in landfills that resemble ancient "ziggurats," bringing offerings to the gods of disposal.[5] While Ammons finds the sacred at the site of egress in our commodity culture, DeLillo presents the supermarket in *White Noise* as the point-of-purchase temple of information, the emporium of signs, in postmodern America. Jack's colleague in the popular culture department, "known officially as American environments" (9), conducts his research in the local supermarket. Murray Siskind exults in the superabundance of information: "This place recharges us spiritually, it prepares us, it's a gateway or pathway. Look how bright. It's full of psychic data. . . . Everything is concealed in symbolism, hidden by veils of mystery and layers of cultural material. But it is psychic data, absolutely. The large doors slide open, they close unbidden. Energy waves, incident radiation. All the letters and numbers are here, . . . all the code words and ceremonial phrases. It is just a question of deciphering, rearranging, peeling off the layers of unspeakability" (37–38).[6] Jack is less impressed by the tantalizing promise of sacred signals than he is disturbed by the ubiquitous static: "I realized the place was awash in noise. The toneless systems, the jangle and skid of carts, the loudspeaker and coffee-making machines, the cries of children. And over it all, or under it all, a dull and unlocatable roar, as of some form of swarming life just outside the range of human apprehension" (36).[7] One thinks of the modest confusion and wonder exhibited by George Bush during the 1991 Presidential election campaign when he first encountered the optical scanner at the checkout lines. The marketing public, already acculturated to barcoded pricing, was scornful of a President so out of touch with the contemporary American environment; it could be said that Bush's detachment—and the faltering economy—cost him the election. Similarly overwhelmed by this environment, Jack hears the cricket-chirping of the "holographic scanners, which decode the binary secret of every item, infallibly. This is the language of waves and radiation, or how the dead speak to the living" (326). He is fearful of his inability to decode the supernatural language of the terminals, to distinguish between the signal and the noise of the supermarket.

What, technically speaking, is the nature of Jack's somewhat bewildering sonic experience in the supermarket? In *Information Theory and Esthetic Perception*, Abraham Moles offers fundamental definitions of the concepts of noise and signal. His text was first published in French in 1958, as DeLillo was completing his degree in communication arts. Moles defines "noise" as "any undesirable signal in the transmission of a message through a channel" (78). One immediately poses the question, "undesirable" to whom? The "cries of children" that are filtered through the many channels of the supermarket audio-system may be undesirable to Jack Gladney, since they are not *his* children, but they are nonetheless meaningful signals in the locator system of the transmitting child and its appropriate receptor parent on an identifiable wavelength. One person's "noise" is another person's "signal." Moles encounters this ambiguity immediately:

> At first sight, it seems that the distinction between "noise" and "signal" is easily made on the basis of the distinction between *order* and *disorder*. A signal appears to be essentially an ordered phenomenon while crackling, or atmospherics, are disordered phenomena, formless blotches on a structured picture or sound. In reality, this morphological distinction is logically inadequate . . . there is *no* absolute structural difference between noise and signal. They are of the same nature. The only difference which can be logically established between them is based exclusively on the concept of *intent* on the part of the transmitter: A *noise is a signal that the sender does not want to transmit.* (78–79; Moles's emphasis)

Postmodern fiction repeatedly engages this problematic of the relation of order and disorder. If there were a "clear" morphological distinction between signal and noise, it would be possible for the receptor to sort the message from its surrounding perturbation with a high degree of reliability. But we are told that signal and noise are of the same nature and so intrinsically indistinguishable. Because the factor of *intent* on which Moles relies is only unambiguously available to the transmitter, there will always exist on the part of the receptor some doubt or indeterminacy as to the form of the message. Moles subsequently addresses the dilemma of the receptor:

> We have assumed a transmitter capable of having objectively definable purposes. But in the case of messages from the environment to the in-

> dividual, the transmitter as an individual does not exist; only the receptor remains. We generalize our definition: A *noise is a sound we do not want to hear*. It is a signal we do not want to receive, one we try to eliminate.
>
> But we have just seen that there is no absolute morphological difference between signal and noise. The desire to eliminate takes effect through a mechanism which selectively obliterates perception of some of the message "noise" which reaches an individual. The individual then has a problem of choice: How will he be guided in selecting message elements to accept or reject? (79; Moles's emphasis)

DeLillo represents the postmodern condition in *White Noise* as the continuous ambiguity of the presence of the "message." Is such a message being transmitted, and if so, how does one know that one has heard it correctly? How does one know which message elements to accept or reject? Jack Gladney experiences the supermarket environment as "awash in noise," but Murray Siskind finds it filled to the rim with "psychic data."[8] How does one know whether the sound one hears is a sound one wants to hear or not? "A voice on the loudspeaker said: 'Kleenex Softique, your truck's blocking the entrance'" (36).

As Jeremy Campbell observes in *Grammatical Man*, noise is the most probable state, when all the parts of the system are mixed up at random; the capacity to convey meaning is improbable (45). Messages are always "immersed in noise, because noise is disorder" (67). The challenge for Claude Shannon, the founder of information theory in 1948, was "how to separate messages from noise, or order from disorder, in a communications system" (67). Such is the problem that the hapless academic Jack Gladney faces; and such is the problem that Don DeLillo set for himself as an American novelist in the 1980s. Moles comments:

> *Noise* thus appears as the *backdrop of the universe*, due to the [entropic] nature of things. The signal must stand out from noise. There is no signal without noise, no matter how little. Noise is the factor of disorder contingent on the intent of the message, which is characterized by some kind of order. It introduces a dialectic, figure-ground, connected with the dialectic, order-disorder, which constitutes the second law of thermodynamics. The general theorem about entropy, "disorder can only increase in an isolated system," amounts to saying that noise can only degrade the orderliness of the message; it cannot increase the particularized information; it *destroys intent*. (85–86; Moles's emphasis)

Gladney is at least partially competent to distinguish simple noises that have readily identifiable sources such as the loudspeaker blaring irrelevancies, children crying wantonly, and shopping carts with spastic wheels squeaking remorselessly. But he is incapable of distinguishing the figure from the ground, either "over" or "under it all," in the "dull and unlocatable roar, as of some form of swarming life just outside the range of human apprehension" (36). Jack is himself perturbed by the nagging suspicion, or perhaps it is fond hope, that this dull roar actually conceals a complex meaning, that it is not merely the static hum of white noise. The more complex the message, the more likely it will be finely interspersed with noise. This "mystery" or complexity is hardly about to reveal itself in a rigid dialect of order and disorder, figure and background. Much abstract art, postmodern fiction, and experimental music have already abandoned the unlikely presentation of such a dialectic. In an interview, DeLillo has called for "the novel that tries to be equal to the complexities and excesses of the culture." He warns, "We have a rich literature. But sometimes it's a literature too ready to be neutralized, to be incorporated into the ambient noise" ("Art of Fiction" 290). Jack Gladney, standing in Paper Products, tries with some consternation to determine whether there can be found within the ambient noise of the supermarket a complex—perhaps transcendental—signal he wishes to hear. DeLillo's fiction naturally situates itself as a communicative art between the orderly figure of pure signal and the disorderly ground of corrupting noise. As a novelist he wishes to represent the full and sometimes maddening cultural complexity of postmodern America without being absorbed by its ambient noise.

At the other end of the auditory spectrum from the unadulterated signal we find white noise. Moles provides the technical definition of white noise as "a more or less continuous noise consist[ing] of erratic repetitions of elementary shocks, recurring with such a large average density as to become indiscernible, yet without any correlation of either amplitude or interval of succession (that is, with the auto-correlation function always zero)" (81). Interestingly, the sound of rain against a windowpane, the monotony of the surf that so unnerved Matthew Arnold in "Dover Beach" to make him iambically cry "Listen! you hear the grating roar / Of pebbles which the waves draw back, and fling, / At their return, up the high strand, / Begin, and cease, and then again begin, / With tremulous cadence slow," or the whistling of a tea kettle at the boil, adequately represent white noise before the advent of electronic technology. But it is at the higher intensities that we associate with the abject failure of the television picture tube or the radio signal that has made white noise a contemporary demon-in-the-machine. Moles states, "By analogy with light, this is what we call a *white noise*, the prototype of perfect, ideal noise. It has no characteristic other than its perfect disorder; white noise is the expression of perfect disorder" (81).

Human experience is comprised of an interpolation of orderly and disorderly activity. Attempts to render human behavior perfectly orderly lead to several forms of unfortunately maniacal dysfunctions.[9] The duly famous episode of Samuel Beckett's *Molloy*, in which the eponymous character first soothes himself with a "sucking stone" and then later becomes obsessed with exhausting a mathematical permutation of sixteen stones circulated between his mouth and four pockets, should serve as a literary example of orderly behavior gone awry.[10] While human behavior fails to maintain a strict orderliness, neither is it capable of exhibiting a perfect disorder or pure randomness. Various forms of chaos, catastrophe, fault, serendipity and accident are possible through human instigation; but the form of perfect disorder actually belongs to the realm of the *not me*. It is for this reason that white noise casts such a spell over the human psyche. Its wholly unpredictable and inscrutable signal inspires a form of awe reserved for that which is entirely alien to our consciousness. The astronomer John D. Barrow offers an intriguing study of the correlation of science and art in *The Artful Universe*. In a chapter devoted to the auditory profiles of noise and music, he shows that because of the entirely random character of white noise, "every sound is completely independent of its predecessors" (232). It fails to stimulate the mind's attention to pattern, a desire that is usually gratified in musical compositions. "Like the spectral mixture that we call white light, white noise is acoustically 'colourless'—equally anonymous, featureless, and unpredictable at all frequencies, and hence at whatever speed it is played" (232–33). The absence of differentiation—the auto-correlation function is always zero—suggests the sonic equivalent of Deleuze and Guattari's conception of "smooth space," unlimited in any direction and self-similar in every dimension.[11] This function of white noise would preclude the mind from concentration for as long as the sound commanded attention. Barrow observes that "at low intensities, white noise has a soothing effect because of its lack of discernible correlations. Consequently, white-noise machines are marketed to produce restful background 'noise' that resembles the sound of gently breaking ocean waves" (233). Despite Matthew Arnold's complaint, New Age tapes and naturist recordings dote on the calming effects of white noise. The troubled ego's interference is dampened. But at higher intensities the application of white noise has precisely the opposite effect, creating a frenetic distraction whose prolonged application could only be described as torture for the identity. Brain-washing. It seems there must be some point of transition in the intensity or frequency (low hum or high squeal) of white noise corresponding to brain waves in which the salutary and open sound-space closes into a dense and mind-obliterating blizzard.

In addition to its effects on the powers of concentration, one should consider the effect that the anonymous, featureless qualities of white noise can have on the identity of the listener. White noise is a static background against which no figure, pattern, or signal is discernible. This quality is particularly distressful to Jack Gladney. Moles examines "the emergence of form in noise, in the birth of an identifiable character in noise." He asks, "If the sonic material of white noise is formless, what is the minimum 'personality' it must have to assume an identity? What is the minimum of spectral form it must have to attain individuality? This is the problem of 'coloring white noise'" (82). Gladney's fear of death, and the absorption of identity into anonymity that it portends, is cast upon the shores of white noise. Listening to its sound, he hopes to identify "some form of swarming life just outside the range of human apprehension" (36). He is trying to "color white noise." He is looking for the emergent signal against the formless background noise. In one of several meditations on death in the novel, Jack and Babette consider what that state of non-being would resemble:

> "What if death is nothing but sound?"
> "Electrical noise."
> "You hear it forever. Sound all around. How awful."
> "Uniform, white."
> "Sometimes it sweeps over me," she said. "Sometimes it insinuates itself into my mind, little by little. I try to talk to it. 'Not now, Death.'" (198–99)

In information theory, white noise represents a state of maximum entropy. The random motion of the electrons in the circuitry has arrived at the most disorderly state possible. The information content of such a signal is zero. A radio listener whose enjoyment of a news program is interrupted by a transmitter failure would experience the vertiginous plunge to the zero information of crackling atmospherics. Jeremy Campbell explains that it "takes longer to describe the message than to describe the noise, because we know everything about the internal order of the first and nothing about the internal disorder of the second" (34). Jack and Babette have associated their lives with the transmission and reception of intelligible signals. They envision themselves as perhaps no more than complex broadcasts within a system of communications technology. The interruption of their particular signal can only be represented as the state of maximum entropy and zero information that we've termed white noise. They conceive of the dissolution of their identity as the immersion into featureless and undifferentiated sound. They are "Off the Air." Ironically, in imagining this "awful" bolgia of the airwaves as an incessant electrical noise, Jack and Babette have given a new face to the figure of Death in the Information Age. Even as they contemplate the static of this figureless ground, they still cannot resist anthropomorphizing, or "coloring," white noise.

In the final section of the novel, "Dylarama," Jack descends upon the trashy Roadway Motel in Iron City intending to murder brain-addled Willie Mink, who had been exchanging the bootleg psychopharmaceutical Dylar for Babette's sexual favors. Jack's motives are a confused compound of revenge for his wife's unfaithfulness, a compulsion to obtain enough Dylar to assuage his own Fear of Death, and the decision to act upon Murray Siskind's theory that killing first is the best remedy against dying. Carefully attended to, the entire passage describes a descent into the undifferentiated buzz of white noise. The unremarkably named motel announces itself as the place where the Death of the Signal will occur. On the usual marquee "were little plastic letters arranged in slots to spell out a message. The message was: NU MISH BOOT ZUP KO. Gibberish but high-quality gibberish" (305). Interference has already begun to shuffle the signal. Jack's deathward plot[12] threatens to absorb him into the very channel of transmission: "I sensed I was part of a network of structures and channels. I knew the precise nature of events. I was moving closer to things in their actual state as I approached a violence, a smashing intensity. Water fell in drops, surfaces gleamed" (305). It appears to Jack that his actions will result in the instant clarification of the message—the "actual state" of things revealed. "Information rushed toward me, rushed slowly, incrementally" (305). But in fact the illusion will result in the intensity, like rainwater pelting a window, of white noise. Confronting Mink, who disjointedly mutters the familiar phraseology of TV voice-overs, Gladney hears "a noise, faint, monotonous, white" (306) in the room. The closer Gladney comes to murder, the more he becomes aware of the presence of white noise: "Auditory scraps, tatters, whirling specks. A heightened reality. A denseness that was also a transparency. Surfaces gleamed. Water struck the roof in spherical masses, globules, splashing drams. Close to a violence, close to a death" (307). At high intensity the disordered signal of white noise approaches a "denseness" or auditory opacity; at low intensity the same disorder can appear transparent. Gladney hopes to experience a "heightened reality" that is meaningful and transcendental. Surrounded by "White noise everywhere," he himself has become the "white man" (310). Suddenly he "understood the neurochemistry of my brain, the meaning of dreams (the waste material of premonitions). Great stuff everywhere, racing through the room, racing slowly. A richness, a density" (310) Jack's dying presence in the channel is just as likely to reveal the meaninglessness of his reality. As Mink pops his stash of Dylar, "His face appeared at the end of the white room, a white buzz, the inner surface of a sphere" (312). Signal is only distinguished from noise by the presence of *intent*. Jack's "plan" requires that he identify himself to Mink by name and explain his motive for murder. Babette's pseudonym for Mink is Mr. Gray. It is against this gray, nebulous background that Jack wishes to raise the more colorful figure of his message. That message is transmitted from Gladney to Mink in the form of a bullet. When the gun fires "The sound snowballed in the white room, adding on reflected waves." As Mink bleeds from the midsection, Jack claims revelation: "I saw beyond words. I knew what red was, saw it in terms of dominant wavelength, luminance, purity. Mink's pain was beautiful, intense" (312). Just as he, Mink, and the motel room have become totally awash in meaningless noise, Jack believes that he has momentarily succeeding in coloring white noise, giving it a figure, causing a dominant wavelength to reveal itself, achieving a transmission that transcends the channel of words. Unfortunately, he is mistaken.

DeLillo's novel introduces a new and disturbing malady into the lexicon of neurological disorders. As the members of the department of American environments—composed almost exclusively of émigrés from the media capital of the country, New York—converse over lunch, Alfonse Stompanato (one of the few Italian-American ethnic figures in DeLillo's work, and here under the guise of an intellectual don) declares: "we're suffering from brain fade. We need an occasional catastrophe to break up the incessant bombardment of information" (66). Just as the miners in the Pennsylvania coal fields during the Industrial Age suffered from the debilitating effects of black lung disease,[13] so the media-saturated in the Information Age suffer from brains blackened by a pulverized silica of data. "The flow is constant," Alphonse observes to his crew. "Words, pictures, numbers, facts, graphics, statistics, specks, waves, particles, motes" (66). This information lacks the strings of logic or the purposeful syntax that would allow for the sorting and grading of data according to levels of significance, or the synthesizing function that permits assembling the particulate into some larger pattern. Brain fade occurs among the unfortunate Appalachians of the airwaves when the mind is no longer capable of responding to the superabundance of information. In Darwinian terms a sudden and violent change in one facet of the environment has stressed the adaptivity of the species. The overtaxed neurons and synapses of the victims require ever more explosive reports to stimulate a response.[14] "Only a catastrophe gets our attention," claims Stompanato. "This is where California comes in. Mud slides, brush fires, coastal erosion, earthquakes, mass killings, et cetera" (66).

John Barrow verifies our fictional scholar's theory when he shows that *black noise* is the result of a high degree of correlation such that the noise is rendered predictable and hence memorable. "Such processes," he remarks, "seem to describe the statistics of a wide variety of man-made and natural disasters—from earthquakes and floods to stock-market plunges and train crashes. The highly correlated appearance of such catastrophes . . . leave[s] no expectation unfulfilled, while white noises are devoid of any expectations that need to be fulfilled" (233). Rowing across the hundred-year flood stage of uncorrelated information, the victims of media saturation can no longer identify the submerged landmarks of their environment. Symptoms include pervasive anomie, a disproportionate response to the magnitude of events, and loss of appetite for anything other than pre-prepared foods. Like the effects of weightlessness in space, brain fade offers the neurologist the first syndrome of extended exposure to the media ecology.

Throughout *White Noise* DeLillo suggests the intensifying correlation (leading to some yet-to-occur apocalypse) between an environment presented as a complex informational system and an understanding of the brain as a densely wired packet of electrochemical connections charged with processing stimuli. Evaluation of the biosphere and the brain in organic terms has given way to a cybernetic analysis[15] that makes coextensive and convergent the subjective experience of the mind and the objective occurrence of the world-net.[16] Input and output become indistinguishable. The excesses and failures on the informational grid of the environment must have direct consequences for the neurological composure of the mind. Jack exclaims, "Isn't it all a question of brain chemistry, signals going back and forth, electrical energy in the cortex? How do you know whether something is really what you want to do or just some kind of nerve impulse in the brain?" (45). In an interview with Tom LeClair that precedes the composition of *White Noise*, DeLillo complains that the defamiliarization of a "crazed prose" is necessary to punch through the brain fade of an audience "drowning in information and in the mass awareness of things. Everybody seems to know everything. Subjects surface and are totally exhausted in a matter of days or weeks, totally played out by the publishing industry and the broadcast industry" ("Interview with Don DeLillo" 87).

Another symptom of "brain fade" that results from the constant bombardment of broadcast media is the imprinting of commercial brand names and other consumer-related phrases on the unconscious mind. Such insignia and catch-phrases—usually in chant-like groups of three—are scattered as leitmotifs throughout the novel, sometimes represented as emanating from the first-person narration of Jack Gladney, but often enough presented as the free-floating signifiers of a media culture. In flight from the toxic cloud of Nyodene-D, Gladney tenderly bends over his sleeping daughter Steffie, only to hear her utter "two clearly audible words, familiar and elusive at the same time, words that seemed to have a ritual meaning, part of a verbal spell or ecstatic chant. *Toyota Celica*" (155). In place of some intimate expression of a psyche in crisis, Steffie "was only repeating some TV voice. Toyota Corolla, Toyota Celica, Toyota Cressida. Supranational names, computer-generated, more or less universally pronounceable. Part of every child's brain noise, the substatic regions too deep to probe" (155). DeLillo surely intends a critique of the pervasive marketing techniques of multinational corporations in the late-capitalist period. He also satirizes the shallow mysticism that runs through the crystalline New Age. Although the utterance strikes Jack "with the impact of a moment of splendid transcendence" (155),[17] these phrases are no more than sound bytes—noisy spores that have attached themselves to the cortex of the witless consumer. DeLillo explains that these "computer generated" words are "pure chant at the beginning. Then they had to find an object to accommodate the words" ("Art of Fiction" 291). The signifier in the media culture truly precedes the material signified. In this context the brain becomes no more than another programmable circuit, an extension of the electronic network in which power and money flow to the automobile manufacturer or tobacco producer who most successfully disseminates its identifying code. Once the multinationals have penetrated the "substatic region" of your child's brain, another young consumer has been territorialized.

Brain fade due to a superabundance of information manifests itself in the Gladney family in one further manner. At two points in the text the family conversations break down into a series of unrelated and garbled factoids. These exchanges—in which nothing of importance is communicated—are symptomatic of the reflux or overload of information. The first instance riffs like an out-of-tune saxophonist on the rivalry of siblings as they try to correct each other's misremembrances. "*The Long Hot Summer,*" Heinrich said, "happens to be a play by Tennessee Ernie Williams" (recombining elements of popular film, country music, and drama). As Jack observes, "The family is the cradle of the world's misinformation. There must be something in family life that generates factual error. Overcloseness, the noise and heat of being" (81). He sees the family unit as an overloaded circuit in a communications channel. A later conversation also ends with a battery of facts partially recalled from school days and now largely obsolete knowledge in the information age. Babette announces, "The battle of Bunker Hill was really fought on Breed's Hill. Here's one. Latvia, Estonia and Lithuania" (176). Jeremy Campbell observes that "computers are good at swift accurate computation and at storing great masses of information. The brain, on the other hand, is not as efficient as a number cruncher and its memory is often highly fallible" (190). The Gladney family demonstrates the shortcomings of the brain in terms of the storage and retrieval of information. More disturbingly, they appear content to engage one another in Read Only Memory bytes rather than exercising the human capacity for intellectual flexibility and synthetic apprehension. When information is detached from context or motive—without a clear *intent* on the part of the transmitter—it becomes indistinguishable from noise. The dysfunctional family in the Information Age behaves like an overheated and error-prone communications channel.

Waves and Radiation

DeLillo describes the three-part structure of *White Noise* as "an aimless shuffle toward a high-intensity event," followed by "a kind of decline, a purposeful loss of energy" ("Art of Fiction" 286). The first part of the novel conceives the randomness of postmodern America in terms of pervasive and undirected "Energy waves, incident radiation" (37–38).[18] The narration strives for a detached and serendipitous quality. As Jack remarks, "Let's enjoy these aimless days while we can, I told myself, fearing some kind of deft acceleration" (18). Although DeLillo is normally averse to writing "highly plotted novels" ("Art of Fiction" 298), he seems intent on fashioning a narrative that emulates the characteristics of Brownian motion. As remarked earlier, the text is dotted with snippets of TV voice-overs that are unmotivated in their context. For example, "The TV said: 'And other trends that could dramatically impact your portfolio'" (61). These voices have broken free from the usual mode of communication. Who is the (often unseen) announcer? Why is he addressing us? The narrative is in constant threat of being overwhelmed by such incident radiation. The unattended TV blaring in the home writes the collective fiction of postmodernism. Much of the action reported in the first section of the novel seems similarly unmotivated: "Blue jeans tumbled in the dryer" (18). No suggestion of causality need even be considered. Those jeans may be tumbling still.

DeLillo expresses in Jack Gladney an anxiety of technology that is notable among postwar writers. He remarks that "There's a certain equation at work" in *White Noise*. "As technology advances in complexity and scope, fear becomes more primitive" ("Art of Fiction" 286). And in a radio interview with Ray Suarez he comments that "technology seeps into our consciousness. I think it makes us docile even as it makes us guilty. It knows everything about us. And it causes a level of anxiety that is sort of quietly pervasive. I mean, in the technology of consumer fulfillment we enjoy what we have, at least up to the point when we may begin to feel certain misgivings. In the technology of industry, we worry about the damage to the environment. . . . We haven't learned to trust technology. . . . There are some difficulties, there are human anxieties, that can't be satisfied by the most sophisticated technologies" (Interview with Suarez, n.p.). Gladney echoes this sentiment when, observing Heinrich's prematurely receding hairline, he suspects "some kind of gene-piercing substance" may have contaminated the environment: "Man's guilt in history and in the tides of his own blood have been complicated by technology, the daily seeping falsehearted death" (22). The primitivization of fear is easily explained. As chair of Hitler studies, Gladney devotes himself to an analysis of the embodiment of Nazism in Adolf Hitler, carrying a dogeared copy of *Mein Kampf* with him everywhere. At mid-century one could locate the material force of deadly threat; another campus might sport departments devoted similarly to Khrushchev or Mao. But the threats of toxic waste or nuclear radiation pose invisible and powerfully penetrating dangers. The new technology's threats are supersensual and disembodied; so they conjure the fear one finds expressed in the ancient mysteries.

Heinrich, however, is born to the Information Age and scoffs at his father's attachment to material cause. Driving to school, Jack and Heinrich debate whether it is presently raining or not. Heinrich states that "The radio said tonight" (22), deferring the sensory experience of a wet windshield to the authority of electronic mediation. Challenged by Jack to recognize that "Just because it's on the radio doesn't mean we have to suspend belief in our senses," Heinrich responds:

> "Our senses? Our senses are wrong a lot more often than they're right. This has been proved in the laboratory. Don't you know about all those theorems that say nothing is what it seems? There's no past, present or future outside our own mind. The so-called laws of motion are a big hoax. Even sound can trick the mind. Just because you don't hear a sound doesn't mean it's not out there. Dogs can hear it. Other animals. And I'm sure there are sounds even dogs can't hear. But they exist in the air, in waves. Maybe they never stop. High, high, high-pitched. Coming from somewhere." (23)

Heinrich's frontal assault on Newtonian physics and the reliability of empirical observation lacks the coherence of the Panzer blitzkriegs through Western Europe. DeLillo satirizes the confusion that the field theories of the new physics breed in the partially comprehending mind of the layperson. But we can still abstract from Heinrich's flurry of assertions a relative measure of the anxiety instilled in the postmodern sensibility by the paradigm shift in the sciences. In his half-absorbed manner, Heinrich offers a version of the Heisenberg Uncertainty Relation as support for why "nothing is what it seems." The observer is always implicated within the same field as the object observed. Any observations that the observer records will reflect the observer's own disposition. Thus the object will appear to have changed with each instance of observation. Because Heisenberg's principle is based upon his study of particle physics, our senses are not only unreliable at such a level of inquiry, but they are largely incapable of recording observations without instrumental aid. Heinrich argues that the unceasing sound waves with which we should be most concerned are those whose high frequencies are inaudible. He strikes a note of paranoia by suggesting that these supersensual waves and radiation are "Coming from somewhere." If the most essential and the most fearsome transmissions from our environment are beyond the range of our sensory reception, then human experience will have to be mediated by technology. Jack's response is sarcastic but clearly worried: "'First-rate,' I told him. 'A victory for uncertainty, randomness and chaos. Science's finest hour'" (24).

In an episode that prefigures the Airborne Toxic Event of the second part of the novel, the grade school in Blacksmith must be evacuated due to a mysterious contamination that sickens the students and teachers. The school is so replete with possible sources of contamination that nothing escapes the suspicion of investigators: "the ventilating system, the paint or varnish, the foam insulation, the electrical insulation, the cafeteria food, the rays emitted by microcomputers, the asbestos fireproofing, the adhesive on shipping containers, the fumes from the chlorinated pool, or perhaps something deeper, finer-grained, more closely woven into the basic state of things" (35). In fact the substances hazardous to human health are so pervasive that they are indistinguishable from the ground of our existence. As "men in Mylex suits and respirator masks made systematic sweeps of the building with infrared detecting and measuring equipment," they seek to isolate the warning signal of hazardous material against a more benign, finer-grained background. But, Jack reports, because "Mylex is itself a suspect material, the results tended to be ambiguous" (35). Here again the very presence of the observer contaminates the observation process. The infrared detectors reflect the presence of the observer in the field. DeLillo implies that insofar as technology emits the rays, fumes and toxins that imperil us, we cannot rely on the products of that same technology to insulate us from death. The insulation is deadly! The measuring equipment is incapable of distinguishing between the subject and the object of detection, the signal and the noise.

At the radio telescope in Arecibo, Puerto Rico astronomers monitor the reception from distant galaxies, hoping to discover a redundant, organized signal that would indicate intelligent life forms against the background hum of the universe. So far nothing; but the scanning of such distant radio signals suggests that one recognizes advanced life forms by the radiation they emit. In the third part of *White Noise* Heinrich dismisses the family's concerns regarding the localized *hot* spills of "Cancerous solvents from storage tanks, arsenic from smokestacks, radioactive water from power plants" (174). Although these toxic spills from the industrialized sector are dangerous within a limited radius, he argues that the "real issue" lies in the *cool* "radiation that surrounds us every day. Your radio, your TV, your microwave oven, your power lines just outside the door, your radar speed trap on the highway" (174); and now one can add the cellular telephone pressed to the ear. Such radiation is pervasive, and in Heinrich's opinion, far more dangerous in low doses that the authorities are letting on. "Forget spills, fallouts, leakages. It's the things right around you in your own house that'll get you sooner or later. It's the electrical and magnetic fields" (175). The *sanctum sanctorum* of the upper-middle class, the suburban home, which should be a refuge from the miseries that plague the less fortunate or able in late capitalist society, turns out to be ground zero in the irradiated apocalypse. As for those who live near the high-voltage power lines that connect every subdivision, "What makes these people so sad and depressed? Just the *sight* of ugly wires and utility poles? Or does something happen to their brain cells from being exposed to constant rays?" (175). John Barrow suggests that "our nervous system may act as a spectral filter to prevent the brain being swamped with uninteresting white background noise about the world, because that background noise is unlikely to contain information vital for survival" (236). But what if the postmodern American home is so flooded with cool white radiation that the brain's capacity to filter such static is overwhelmed? Or as Heinrich warns, "There are scientific findings. Where do you think all the deformed babies are coming from? Radio and TV, that's where" (175). Our very survival, DeLillo wryly points out, is threatened by the uninteresting, meaningless emanations from these media.

A Period of Chaos

In DeLillo's conception of *White Noise,* the "aimless shuffle" through the noisy terrain of "Waves and Radiation" moves toward a "high-intensity event" in "The Airborne Toxic Event," the briefest and most dramatic of the novel's three sections ("Art of Fiction" 286). The family's response to the derailing of a tanker car leaking a cloud of deadly Nyodene D—the "byproducts of the manufacture of insecticide. The original stuff kills roaches, the byproducts kill everything left over" (131)—varies predictably with their established characterization. Jack, who reveals his fragile and suspect authority as the non-German-speaking chair of Hitler studies, assumes leadership of the family evacuation with "an air of weary decisiveness" (115) that more closely resembles reluctance. He complains that "I'm the head of a department. I don't see myself fleeing an airborne toxic event. That's for people who live in mobile homes out in the scrubby parts of the county" (117). Disaster victim is not part of his job description. His professional status and suburban home prove to be insufficient insulation from what he views as an inconvenient disruption of service. Ironically, he is the only family member exposed to the deadly cloud while pumping gas at a service station along the evacuation route. Heinrich, however, demonstrates his messianic complex. He is the most vigilant and knowledgeable member of the family. He enthralls an audience of fellow refugees at the Red Cross shelter with his analysis of the event in the manner of the young Jesus preaching in the temple in Jerusalem. His language combines the folksy humor and textbook presentation of the expert commentator on the nightly newscast. He is Moses and Dan Rather speaking to the gathered multitude "in the name of mischance, dread and random disaster" (131). The remaining children demonstrate their impressibility as they exhibit each of the symptoms of exposure as reported by the radio.

The radio reports that update the status of the toxic cloud from "feathery plume" to "black billowing cloud" and finally to the official jargon of "airborne toxic event" mediate the family's conception of the disaster. Although precipitated by a transportation accident, the disaster is quickly staged by the media and federal authorities as if it were the Hollywood premier for a technicolor film.[19] Jack describes the scene as "one of urgency and operatic chaos. Floodlights swept across the switching yard. Army helicopters hovered at various points, shining additional lights down on the scene. Colored lights from police cruisers crisscrossed these wider beams" (115). It's no wonder that the drab Gladneys assume more extravagant roles during the incident. As if to emphasize the media-controlled presentation of this man-made chaos, the disaster response team operates under the sobriquet of SIMUVAC. They regard the event as an opportunity to rehearse for a simulated evacuation: "You have to make allowances for the fact that everything we see tonight is real" (139). Simulated relief at least provides the appearance of order. "What people in an exodus fear most immediately," Jack muses, "is that those in positions of authority will long since have fled, leaving us in charge of our own chaos" (120). Operas from Giuseppe Verdi's *Otello* to John Adams's *The Death of Klinghofer* have always represented human tragedy rather than naturally induced disaster. The toxic cocktail spilled on Blacksmith fits that billing. The domestic tranquillity of a small college town is ironically interrupted by the massive release of an insecticide intended for home use.

Several critics have mistakenly identified the black cloud of Nyodene Derivative and the white noise of background radiation as comparable phenomena, perhaps because they both portend death in the novel. But as previously discussed, white noise is an example of pure randomness; it is the "expression of perfect disorder" (Moles 81) in a system. No correlation or patterning is evident. The black cloud that threatens Blacksmith—though a man-made irruption—is more properly an example of chaotic behavior. Like the black noise that describes "the statistics of a wide variety of man-made and natural disasters" (Barrow 233), such chaotic disturbances display a much higher degree of correlation or patterning. Barrow remarks that the "highly correlated appearance of such catastrophes could be taken as a basis for the old adage that 'accidents always come in threes.'" Black noises "leave no expectation unfulfilled, while white noises are devoid of any expectations that need to be fulfilled" (233). The mindless wandering through white noise is punctuated by a sudden correlation of conditions—track condition, freight schedule, and wind direction—that bring catastrophe down upon Blacksmith. The disturbance is not predictable, but it is not unorganized either. The black cloud is a highly correlated phenomenon that remains beyond human control. Jack describes the experience of chaos as an encounter with the sublime:

> The enormous dark mass . . . was a terrible thing to see, so close, so low, packed with chlorides, benzines, phenols, hydrocarbons, or whatever the precise toxic content. But it was also spectacular, part of the grandness of a sweeping event. . . . Our fear was accompanied by a sense of awe that bordered on the religious. It is surely possible to be awed by the thing that threatens your life, to see it as a cosmic force, so much larger than yourself, more powerful, created by elemental and willful rhythms. This was a death made in the laboratory, defined and measurable, but we thought of it at the time in a simple and primitive way, as some seasonal perversity of the earth like a flood or tornado, something not subject to control. (127)

Jack's training as an academic in the Western tradition allows him to appreciate certain qualities in the black cloud not comprehended by most of those among the "tragic army of the dispossessed" (127). He asserts a correlation between the nature of chaos and the experience of the sublime, between the phenomenon and its apprehension. The chaotic behavior of the cloud is at once self-organizing, "created by elemental and willful rhythms," and unpredictable, "not subject to control." It displays intensifying patterns (as does a flood or tornado or stock market collapse) without exact repetition. Similarly, the sublime as an aesthetic category combines the pleasing grandeur of a large-scale correlated phenomenon with a sense of awe or fear before a force that is beyond human control.[20]

In Part III, "Dylarama," Jack encounters the chaos that has infiltrated his life on a smaller scale than the apocalyptic cloud that looms over Blacksmith. Searching through the trash compactor for the remnants of Babette's stash of fear-suppressing Dylar, he deconstructs the anxiety that he feels for much that is disorderly and uncontrollable in the domestic sphere. In the crushed density of the compactor bag Jack attempts to isolate a single meaningful component in his life. "The bottles were broken, the cartons flat. Product colors were undiminished in brightness and intensity" (258). He is searching for a signal of similar brightness and intensity in the noise-disrupted and entropically reduced waste of postmodern living. "The compressed bulk sat there like an ironic modern sculpture, massive, squat, mocking" (258–59). With a snide swipe at art criticism, DeLillo suggests that art (in whatever medium) only reluctantly surrenders denotational meaning. Sculpture formed from the waste of automobile bumpers, steel girders, cogs and flywheels remains indistinguishable in its complexity from the ground of existence. As if to illustrate his critical consciousness, Jack roots from the muck and detritus a form of particular interest: "There was a long piece of twine that contained a series of knots and loops. It seemed at first a random construction. Looking more closely I thought I detected a complex relationship between the size of the loops, the degree of the knots (single or double) and the intervals between knots with loops and freestanding knots. Some kind of occult geometry or symbolic festoon of obsessions" (259). Critics of DeLillo have taken this looped and knotted string as symbolic of his own construction of novels.[21] Jack's analysis reads like a compressed text on complexity theory. Although the string appears to be a "random construction" without any correlation in the loops and knots, further study reveals a "complex relationship" indicative of design or intention. One might counter, however, that the discovery of such "occult geometry" reveals nothing more than the mind's need to recognize patterns. The indeterminacy is almost unbearable: is the string a meaningless form that may not be parsed; is it a spontaneously emergent form of complexity; or is it merely a frustrating reflection of one's own "festoon of obsessions"? The string represents DeLillo's penchant for fiction that is "equal to the complexities" ("Art of Fiction" 290) of the world, writing which "involves a centering and a narrowing down, an intense convergence. An obsessed person is an automatic piece of fiction" ("Interview with Don DeLillo" 88).

Murray Jay Siskind is just such an "automatic piece of fiction." He practices total immersion therapy in the American environment of supermarkets, tabloid journalism, and run-down rooming houses near the insane asylum. "I'm totally captivated and intrigued" (10), he says, by the superfluity of American popular culture. With the obsessive zeal of a motivational speaker, Murray coaches the more skeptical and frequently overwhelmed Gladney. Jack believes that his exposure to Nyodene D—with its lifespan of thirty years in the human body (141)—will kill him; but at age 51 he won't know for possibly fifteen years whether symptoms will manifest themselves. He acknowledges that "Death has entered" (141), but he suffers in prolonged indeterminacy. He cannot say for certain whether the complexly knotted string or the data on his exposure represents a meaningful pattern or an insignificant correlation. He cannot isolate the signal from the noise. Murray prescribes the assertion of plot and design against the indiscriminate welter of existence, in effect urging Jack to become the sender of the message rather than the receiver: "We start our lives in chaos, in babble. As we surge up into the world, we try to devise a shape, a plan. There is dignity in this. Your whole life is a plot, a scheme, a diagram. It is a failed scheme but that's not the point. To plot is to affirm life, to seek shape and control" (291–92). In art, politics, and personal development the solution to chaos and drift is a good linear plot. "Are you sure?" Jack replies. Despite the fact that Murray is merely "*talking* theory" (292), Jack initiates his plan to kill Mr. Gray, the indistinguishable and composite figure of a media-saturated society,[22] in an effort to formalize and transmit his own life-story.

In stalking Mr. Gray, Jack reiterates his mantra, "This was part of my plan. My plan was this" (312): to cause his victim a slow and agonizing death by shooting him three times in the midsection—exhausting the ammunition in the small-caliber handgun he obtains from his father-in-law—and then placing the gun wiped clean in his victim's hand to suggest a bizarre sort of suicide. The act affords Jack a brief epiphany, as when a twist of the antennae momentarily brings the TV picture tube to a clear image before it returns to snow and fritz. "I knew who I was in the network of meanings. . . . I saw things new" (312). The act of self-determination allows Jack to tune in his own signal against the background noise of death and disorder, but also to perceive the more extensive connections in the network of communications that had previously been recondite. Unfortunately Jack's scheme fails when he neglects to fire the third round, which Willie Mink perhaps reflexively discharges into Jack's wrist. Loss of signal follows immediately:

> The world collapsed inward, all those vivid textures and connections buried in mounds of ordinary stuff. I was disappointed. Hurt, stunned and disappointed. What had happened to the higher plane of energy in which I'd carried out my scheme? The pain was searing. Blood covered my forearm, wrist and hand. I staggered back, moaning, watching blood drip from the tips of my fingers. I was troubled and confused. Colored dots appeared at the edge of my field of vision. Familiar little dancing specks. The extra dimensions, the super perceptions, were reduced to visual clutter, a whirling miscellany, meaningless. (313)

Jack's visionary experience—the complex connections and coherent images—suddenly ionizes. The slipped bonds of meaning and the deterioration of perception are described in terms of a transmission broken up by interference. The "colored dots" of trauma—as blood rushes to the brain—are noise that speckles the screen as the signal fails. Jack reenters the "whirling miscellany" of chaos. The meaningless disorder of white noise looms. But the field of high-intensity energy and dense nexus of communication through which Jack's receptive consciousness passes may have been generated by the very chaos that so dismays him. The "mounds of ordinary stuff" are undifferentiated like the components of a system at equilibrium. But in a state far from equilibrium—which seems to describe Jack's psychological condition—the prospects for the revelation of deeply complex associations actually improve.

Heat Death

The trajectory of *White Noise* moves from the randomness of Part I to the peak of chaotic intensity in Part II and finally subsides in what DeLillo describes as "a kind of decline, a purposeful loss of energy" ("Art of Fiction" 286) in Part III. This concluding section of the novel represents an increase in the entropy of the system, a return to the state of greatest probability. The town of Blacksmith disappointingly reveals no significant change from its ordeal—as in the natural disasters of flood, fire, or earthquake—beyond the chemically enhanced colors of a "postmodern sunset" (227). Holding up an example of the offending television set, one evacuee complains that the incident receives no network coverage, "No film footage, no live report. Does this kind of thing happen so often that nobody cares anymore? . . . What exactly has to happen before they stick microphones in our faces and hound us to the doorsteps of our homes, camping out on our lawns, creating the usual media circus?" (161–62). Far away from the convulsions of a "polestar metropolis" (177), small town life returns to normal.

The distinction drawn between the fulsome coverage devoted to urban disasters and the attention paid to an apparently insignificant college town serves to illustrate the two related definitions of entropy. At the beginning of the novel Murray Siskind—a New York émigré—postulates an equation between cities and thermodynamic entropy:

> "Heat. This is what cities mean to me. You get off the train and walk out of the station and you are hit with the full blast. The heat of air, traffic and people. The heat of food and sex. The heat of tall buildings. The heat that floats out of the subways and the tunnels. It's always fifteen degrees hotter in the cities. Heat rises from the sidewalks and falls from the poisoned sky. The buses breathe heat. Heat emanates from crowds of shoppers and office workers. The entire infrastructure is based on heat, desperately uses up heat, breeds more heat. The eventual heat death of the universe that scientists love to talk about is already well underway and you can feel it happening all around you in any large or medium-sized city. Heat and wetness." (10)

Siskind conceptualizes the city in thermodynamic terms as an engine. As the engine performs work it dissipates energy in the form of heat. Eventually—on the assumption that the city is an isolated system—all of the energy available to do work will have dissipated, and the system will reach a state of maximum entropy. At maximum entropy the city will no longer be capable of any work and the temperature gradient will have fallen so that the city is no longer distinguishable in temperature from its surroundings—a layperson's version of "heat death." But in the meantime a metropolis like New York, along with its denizens, is hot and sweaty.

The predicament of Blacksmith pertains more closely to informational than to thermodynamic entropy. In general the industrialized cities are hotter than college towns. Information theory is a cool science. Even Hitler studies, which attends to the dissemination of the Hitlerian image of absolute authority, was founded on "a cold bright day with intermittent winds out of the east" (4). Jeremy Campbell explains the relationship of entropy and information: "When a system is orderly, and therefore improbable, when it is low in entropy and rich in a structure on the macroscopic scale, more can be known about that system than when it is disorderly and high in entropy. When it is at equilibrium, the state of greatest change beneath the visible surface on the microscopic scale, but of the greatest 'sameness' as seen by the human observer, we have the least possible knowledge of how the parts of the system are arranged, of where each one is and what it is doing" (43). There is thus a relationship between an increase in entropy and a decrease in knowledge. Ludwig Boltzman first suggested in 1894 that entropy was related to "missing information" (Campbell 44). As a disaster site Blacksmith is a cool rather than a hot spot. The residents experience high entropy due to their uncertainty and lack of information after the dissipation of the cloud of Nyodene D. In the final chapter of the novel, Jack observes that "we don't know whether we are watching in wonder or dread, we don't know what we are watching or what it means, we don't know whether it is permanent, a level of experience to which we will gradually adjust, into which our uncertainty will eventually be absorbed, or just some atmospheric weirdness, soon to pass" (324–25). Although the men in Mylex suits continue to gather data, the residents of Blacksmith pass into a highly entropic postmodern condition described by a vast quantity of missing information.

Data

As the pretentiously dubbed J. A. K. Gladney, Jack has founded his professional career on an analysis of the "power principle" (258) as it is expressed in the mystique of Adolf Hitler. Jack's course of study in "the continuing mass appeal of fascist tyranny" (25) relies heavily on the presentation of a charismatic figure, to the extent that his adoption of impenetrably dark glasses and his tendency to "loom" over his gathered students confuses the method of presentation with the object of study.[23] Authority becomes embodied in a single gifted individual. The power asserted by this personality over the adoring masses reaches an unquantifiable level. The curriculum of Advanced Nazism resembles nothing so much as a seminar uncritically devoted to the totalizing, centering ideology of modernism. But a traumatic transformation occurs in Jack's concept of the locus of authority as a result of his exposure to Nyodene D. The chair of the department of Hitler studies must adopt a "servile and fawning" disposition toward a mere foot soldier in the SIMUVAC army because, he says, "I wanted this man on my side. He had access to data" (139). The harbinger of this new paradigm is a particularly inarticulate weekend warrior spouting the jargon of "insertion curve" and "probability excess" (139) in a jumble of sentence fragments and dangling prepositions. The technician is hardly the sort of communicator who would inspire a mass audience. Unfortunately, while Jack in his Ivory Tower lectures to his classes on the power principle, authority has shifted to those who control data. In the Information Age "access to data" translates directly into authority, power, and virtually unlimited funding.

Jack tries to maintain his cultivated persona of "an impassive man" and "counteract the passage of computerized dots that registered my life and death" (140). But data-collecting instruments only quantify the quantifiable; they are insensitive to anything so ineffable as personal charisma. Jack is swiftly digitalized: "You're generating big numbers" (140). The technician naturally disavows responsibility for the results: "It's what we call a massive data-base tally. Gladney, J. A. K. I punch in the name, the substance, the exposure time and then I tap into your computer history. Your genetics, your personals, your medicals, your psychologicals, your police-and-hospitals. It comes back pulsing stars. This doesn't mean anything is going to happen to you as such, at least not today or tomorrow. It just means you are the sum total of your data. No man escapes that" (141). The inarguable judgment of the data is devastatingly impersonal. Identity is reduced to a composite of fields. These fields produce a "sum total" that, like a pointillist painting viewed too closely, can only suggest the holistic figure. In data processing the component parts speak more loudly than the unified subject. Likewise authority in the information system tends to be decentered. Gladney receives what amounts to a deferred death sentence, but who pronounces it? N. Katherine Hayles argues that "contemporary technology, especially informatics, has given us the sense that we transcend [biological, cultural, historical] limitations and live a disembodied, free-floating existence made possible in part by the near-instantaneous transfer of information" ("Postmodern Parataxis" 394). When Babette appears unexpectedly on the local cable station, Jack reacts with funereal grief: "What was she doing there, in black and white, framed in formal borders? Was she dead, missing, disembodied? Was this her spirit, her secret self, some two-dimensional facsimile released by the power of technology, set free to glide through wavebands, through energy levels, pausing to say goodbye to us from the fluorescent screen?" (*White Noise* 104). Such disembodied existence may be liberating for some in the media age, but it also signifies a stunning shift in the locus of authority and power to an impersonal, decentered, and transferable non-entity that doesn't so much loom over its victims as engulf them.

By its definition as factual information that has been presented in a form suitable for analysis or processing, data can be distinguished from disruptive noise. Data should be useful and accurate information. But much of the data presented in *White Noise* seems incoherent and resistant to analysis. On a visit to the Gladney home, Murray Siskind observes that "There were huge amounts of data flowing through the house, waiting to be analyzed" (101). The television attended or unattended; "the otherworldly babble of the American family" (101) transmitted simultaneously on multiple frequencies; autonomous or group traffic patterns through the house; utility and entertainment systems and their relative importance. Murray's exhaustive research into the cultural matrix of the American family may eventually find a publisher, but he has little to report during the course of the novel. For the most part the data remains uncaptured or "free." The Gladneys are immersed in an infinitely extensive, open data flow. If the superabundance of data doesn't tell us what we want to know, or cannot be presented in a form that we can apprehend, how is it distinguishable from noise? Is superfluous or useless data still information? An open, decentered, uncontrolled data field poses notable adaptive problems for the Gladneys. They appear insensate as they perform their everyday obligations because a total appraisal of the cultural codes at play in any given moment would have a paralytic effect on their next action. But as Jack observes, "We certainly didn't need to face each other across a table as we ate, building a subtle and complex cross-network of signals and codes" (231). The generation and reception of such complex data fields would appear to be an innate human characteristic—wired into the genome—not requiring the conscious recognition of their presence.

Limitless data flowing through an open field may be as incomprehensible as it is benign; low-level radiation passes through the body without affecting its systems. But the Gladneys, like all American families, interact with several highly controlled and regulated systems from which they derive a measure of security. Jack's withdrawal of cash and balance check at an automated teller machine elicits a satisfactory response:

> The system had blessed my life. I felt its support and approval. The system hardware, the mainframe sitting in a locked room in some distant city. What a pleasing interaction. I sensed that something of deep personal value, but not money, not that at all, had been authenticated and confirmed. A deranged person was escorted from the bank by two armed guards. The system was invisible, which made it all the more impressive, all the more disquieting to deal with. But we were in accord, at least for now. The networks, the circuits, the streams, the harmonies. (46)

The successful interaction with a small sector of cyberspace that happens to be heavily controlled provides Jack with a modicum of benediction in his life. He is well aware that such a closed system might reject his bankcard, or his secret code, and thus render him inauthentic. He is vulnerable to the whims of the system, but it is far more predictable and reassuring than the "whirling miscellany" of data to which he would be subject beyond its bounds. Yet here too Jack must act in a suprarational manner, assuming an almost mystical belief that he is in accord with the invisible streams of the financial world's data field. Closed systems struggle to purge themselves of noise. The expulsion of the "deranged man" that interrupts Jack's karmic transaction exemplifies the system's self-regulating elimination of noise in the circuit. But if open systems lack structured reliability, closed systems never succeed in completely purging themselves of noise. The system requires of its components a limited identification in exchange for the security of interacting within the system. When Jack receives his new automated banking card, he is reminded, "You cannot access your account unless your code is entered properly. Know your code. Reveal your code to no one. Only your code allows you to enter the system" (295). The reminder strikes an insincere note in suggesting that the system is solicitously concerned with the individual's personal security. In fact one is being disciplined to the security procedures of the system to protect the system from loss or damage.[24] The closed data field, which covets its secure, coherent, and information-rich structure, conditions the user to the limited terms of interaction. Only some approved information may be withdrawn; and the user leaves only a trace or datum of his presence in the system. In an open data field, whose diffuse structure is relatively information-poor, the user has almost unlimited access to data and may contribute data more freely. Jack moves with trepidation before the closed system and with perplexity before the open system.

End Transmission

Among the detached phrases that float through the text, none loom more ominously than the triad of "Random Access Memory, Acquired Immune Deficiency Syndrome, Mutual Assured Destruction" (303); or as they are more commonly known by their acronyms, RAM, AIDS, MAD. What beyond the incantatory similarity of their shorthand monikers brings these three expressions together? The principle of computer data storage and retrieval; a fatal virus whose genetic coding is capable of rapid, defensive mutation; a Cold War policy of nuclear deterrence so insane that its acronym bespeaks all. The associative logic suggests that we have already entered a post-apocalyptic period. Busily warding off biotoxic invaders and nuclear annihilation, no one pays heed to the rippling distortion of the informatics system caused by an enormous data blast. The most insidious device of the nuclear age was the neutron bomb, whose radiation could decimate populations while causing minimal damage to property and infrastructure. The citizens of Blacksmith suffer the devastating effects of a neuronic bomb whose magnetic storm, electrical transmissions, and computerized pulses distort and erase all cognizance of the war itself. One day they arrive to find that the "supermarket shelves have been rearranged. . . . They walk in a fragmented trance, stop and go, clusters of well-dressed figures frozen in the aisles, trying to figure out the pattern, discern the underlying logic" (325). Were they not already victims of a successfully waged campaign of cognitive distortion, they would recognize the shift from a condition of order and improbability in neatly categorized products to one of disorder and probability in which "the condiments are scattered" through the shelves (326). DeLillo's *White Noise* tells of a communications Armageddon that has swept through America, the early warnings of which were lost in the ambient roar of postmodern culture.

6

The Perfect Game

Dynamic Equilibrium and the Bifurcation Point in Robert Coover's *The Universal Baseball Association*

To say that the world is meaningless would mean that we cannot detect any rules, that we don't understand the past and that we cannot predict the future; it would mean total contingency, which seems incompatible with even the precarious existence of an individual conscience. To say that the world is meaningful would signify, if this meaning were perfectly understood, that the past and the future are open before us like a book. The truth is somewhere in between, that is to say that we can figure out partial, limited meanings that allow us to function in certain circumstances, whereas in others we are left powerless.

—Ivar Ekeland, *The Broken Dice*

"Even nonpattern eventually betrays a secret system," Dr. Peloris explained confidently to all present, "but so far that of our subject, which seems largely instinctual, is simply not apparent."

—Robert Coover, *Pricksongs and Descants*

Strat-O-Matic® board games that simulate the play of real-life professional athletes in sports such as baseball, football, and hockey were introduced in 1960. Each game consists of a set of three dice—to increase the number of combinations and the complexity of the possible outcomes—and a series of statistically accurate performance charts for the teams, the athletes on their roster, and various game situations. Players of these board games assume the role of the team's manager, setting the lineup, making substitutions, and calling plays. Rather than identify with an individual athlete, the player takes responsibility for the strategy of the game. As the sales pitch on the box cover declares, "With Strat-O-Matic you are in control!" Any reader of Robert Coover's novel, *The Universal Baseball Association, Inc., J. Henry Waugh, Prop.* (1968), who was ever devoted to Strat-O-Matic® Baseball would legitimately conclude that the board game was a source for the game-in-the-novel.[1] Although Waugh's teams and their personnel are his fictions (not to be confused with so-called "fantasy leagues" that depend on the performance of professional athletes in actual contests), his Universal Baseball Association likewise relies on a set of three dice and elaborate performance charts for the skill levels of each player and the routine or extraordinary occurrences that arise. As league proprietor, Waugh strategizes and controls his teams, even as he is subject to the random throws of the dice and the laws of probability. Coover examines the ludic impulse in all human beings,[2] but he finds there no trivial pursuit or mere sporting entertainment. Waugh's league is a carefully contrived dynamical system whose survival or collapse depends upon complex equilibria between error and perfection, power and control, chance and design. In the 1990s Strat-O-Matic® entered the electronic gaming environment, easily adapting its products to the random-number generation and calculating abilities of the computer. For their part fans have created a number of websites devoted to interleague play. Coover, whose short fictions "Morris in Chains," "The Panel Game," and "The Babysitter" in *Pricksongs and Descants* (1969) were exemplary of the Barthesian *scriptible* text, has more recently paid critical attention to nonlinear interactive fiction and hypertext.[3] Espen Aarseth distinguishes between an "ergodic literature" typified by popular adventure games such as *Doom* and *Myst* that demand that the reader-player take actions that result in altered outcomes, and hypertext fictions such as Stuart Moulthrop's *Victory Garden* (1991) whose readers are "limited to metasemiotic exploration" (*Cybertext* 48–49). Though Coover himself has not turned to the writing of hyperfiction, his analysis of hyperfiction embraces both modes of reader-as-player and reader-as-explorer. He is intrigued by the polyvocality and graphical elements of these texts, and by their "effective use of formal documents not typically used in fictions— statistical charts . . . photographs, baseball cards and box scores, dictionary entries, . . . board games" ("End of Books" 25). In these works and in his own fiction Coover finds the explicit introduction of design.

"Turning back to design"

In an interview with Frank Gado, Coover observes that during the Enlightenment there was "a shift from a Platonic notion of the world—the sense of the microcosm as an imitation of the macrocosm and that there was indeed a perfect order of which we could perceive only an imperfect illusion—toward an Aristotelian attitude which, instead of attempting a grand comprehensive view of the whole, looked at each particular subject matter and asked what was true about it." The shift from an idealist to an empiricist approach to the universe served to "isolate our scientific activities, each from the other, as well as to isolate scientific from literary activity. We began to look at the world as it 'really was'—and to define that 'really was' in terms of the specific. It marked the beginning of realism" ("Robert Coover" 143). The increasing specificity of human inquiry appeared to make abstractions that pertained in multiple disciplines both inadequate and invalid. Coover holds the conviction, shared by John Barth in "The Literature of Exhaustion,"[4] that in the latter half of the twentieth century "we have come to the end of a tradition. I don't mean that we have come to the end of the novel or of fictional forms, but that our ways of looking at the world and of adjusting to it through fictions are changing." Thus, "our basic assumptions about the universe have been altered" once again ("Robert Coover" 142). Jean-François Lyotard defines postmodernism as the expression of incredulity toward master narratives such as rationalism, humanism, and scientific progress (xxiv). In a postmodern condition characterized by uncertainty, pervasive irony, and randomness that diminishes human control, Coover seeks recourse in "turning back to design."[5] The designs in his fiction "have a certain beauty, and now a potential for irony exists in them" ("Robert Coover" 143) since they are the products of an incredulous age. They are empty—because ironic—but still useful abstractions. There is beauty in the perfect game, even though it makes no claim to timelessness. Coover explores the possibilities inherent in these arbitrary orders. "It is easier for me to express the ironies of our condition by the manipulation of Platonic forms than by imitation of the Aristotelian" ("Robert Coover" 143–44), through metafictional propositions rather than realism.

Coover's search for adequate designs to express the change in humanity's relationship to the universe runs through mythopoeic and fabulist figures, literary forms, and scientific paradigms. Any answers to the fundamental questions of existence, he reasons, must be provisional: "In part because individual human existence is so brief, in part because each single instant of the world is so impossibly complex, we cannot accumulate all the data needed for a complete, objective statement. To hope to behave as though this were possible is to invite paralysis through crushing despair. And so we fabricate; we invent constellations that permit an illusion of order to enable us to get from here to there. . . . Thus, in a sense, we are all creating fictions all the time, out of necessity. We constantly test them against the experience of life" ("Robert Coover" 152). Roland Barthes has said that "the work of art is what man wrests from chance" (*Critical Essays* 217). The fabrication of order, the search for pattern in complexity—even if that pattern fails to provide a complete account—sustains art and life.

The hypothetical aspect of Coover's fabricated orders, the need to continuously test their usefulness as a descriptor of the world, presses an analogy with scientific experiment. Gado suggests that while "art responds to change inaugurated by scientific or technological advances," it may also be true that "the poet anticipates the development of new scientific theory" ("Robert Coover" 153). Coover agrees with the proposition that literary and scientific method should be capable of a reciprocal influence, that revolutionary concepts and experimental procedures in one discipline can be provocative in the other. He challenges the isolation of scientific from literary activity, and the assumption of the primacy of discoveries in the physical sciences which fiction can only imitate or emulate. He cautions that "it's not necessary" for revolutionary concepts in the one discipline to "become translated into the other. Essentially, both the new scientific and the new aesthetic concepts emerge because the old ones, having rigidified in forms that have lost contact with the on-goingness of the world, have become impotent" (153). Coover does not concern himself with whether scientific or aesthetic concepts lead or follow in revealing the shape of the universe, but rather that each arrives at homologous and comparable forms that are valid when tested experimentally against the nature of things. Imaginative constructs may aid in the proposition of elementary particles, as when Murray Gell-Mann borrows the neologism "quark" from James Joyce's *Finnegans Wake;* the complementarity theory of light as both wave and particle suggests a paradoxical element in nature to which literature has long been receptive; and much scientific discourse resorts to metaphor in order to describe the unknowable—that which cannot be directly experienced—in terms of the familiar.

The arbitrary designs of literature and the hypotheses of science are both subject to displacement by more apt formulations as understanding changes. Neither is a fixed universal. Coover says, "All of us today are keenly aware that we are undergoing a radical shift in sensibilities. We are no longer convinced of the *nature* of things, of design as justification. Everything seems itself random" ("Robert Coover" 153). In *Fiction in the Quantum Universe,* Susan Strehle argues that "Coover's fiction specifically engages the shift, crucial to recent scientific understandings of the way events occur, from causality to probability" (68). She convincingly demonstrates that Coover's "plots often address the irruption of chance in the lives of characters who have believed in causality" (69). In rejecting the conviction of Newtonian physics that with sufficient information and a set of natural laws all events are predictable, Coover relinquishes the notion of "design as justification," as a teleological determinism that unifies his fictive world and its events. Coover's gaming acknowledges the uncertainty principle in quantum theory that, as Strehle notes, "provides a statistical description of the probability of group behavior but denies that individual events can be predicted" (68). But neither does he consign his fiction to a random series of interactions. Nor would a physicist conducting experiments at a linear accelerator have no expectation of results from a collision of subatomic particles. Thus Coover proposes, "Under these conditions of arbitrariness, the artistic impulse is directed toward putting the random parts together in any order which provides a pattern for living" ("Robert Coover" 153). His work expresses probability even as it formulates a design that provides a suitable understanding of its function in our lives.

Game Theory

The reader of *The Universal Baseball Association* might be said to emulate a "Research Specialist in the Etiology of Homo Ludens" (233). What are the origins of the gaming instinct in humans? What fears or necessities drive humans to the invention of games? Why are some games sustained (despite rule changes and cultural transformations) while others fall into disuse? A game is the expression of the ordering impulse of human consciousness in contest with the patternless series of events. Games such as solitaire or backgammon demand that the players achieve patterns that are regular and inflexible; striving for the completed pattern (no cards remaining at hand, or disks unplayed) against the demanding limits of the rules provides the challenge and the satisfaction. In such games chance will outweigh the player's skill or strategy; no amount of cleverness, short of cheating, can overcome an infelicitous turn of the cards or roll of the dice. Other more complicated games like Mah Jongg,[6] contract Bridge, or Scrabble demand that the players be continuously aware of patterns that are capable of evolving into other forms. Games of strategy represent the evaluation of resources in the face of threats (averted) or opportunities (seized). End-stopped games with a prescribed pattern of completion usually depend upon the opposition of strict rules and the introduction of aleatory elements (shuffling of cards, casting of dice). Distinct from so-called games of chance, there are games of strategy whose final positions are various, in which patterns rarely repeat themselves (the "solution" or final patterning of one instance of the game is almost never that of another), and the player must constantly adapt to the potentialities of the board. Although chance may figure in the complexity of such games, it is not the prevailing factor. Games of strategy are an exercise of the fabricative human mind; they are authorial.

In his public life as a menial accountant for the firm of Dunkelmann, Zauber & Zifferblatt, Henry Waugh's obligations as bookkeeper are monotonously repetitive and meaningless. Henry suffers from "Double entry fatigue. Cancer of the old intangibles, Ziff baby. Wasting assets" (30). At age fifty-six (coincidentally, "this was Year LVI" of Waugh's Association: "he and the Association were the same age, though of course their 'years' were reckoned differently. He saw two time lines crossing in space at a point marked '56.' Was it the vital moment? Silly idea." [49–50]; Damon's father, Brock Rutherford, is fifty-six years old [65]; there are 56 number combinations of the three dice [20]), he has entered a phase of irreversible physical decline not dissimilar from the capital collapse of the firm he's auditing, Meo Roth's Skylight Protection Company (which seems a singularly unpromising trade): "Old Meo Roth was reeling toward the ruin level. 'Join the company,' Henry said" (135). Henry's predicament recalls that of the eponymous anti-hero of B. S. Johnson's *Christie Malry's Own Double-Entry* (1973). As a petty bank clerk Christie Malry feels understandably exploited by the capitalist system; he justifies his recourse to anarchism through the application of "the sublime symmetry of Double-Entry" (23). Malry applies the First Golden Rule to his case: "*Every Debit* [for an offense received] *must have its Credit* [for an offense given]" (24). The double-entry method of accounts is a zero-sum system: credits and debits should offset one another. The diegetic author in Johnson's novel (which is organized according to "Reckonings") reminds us that "the first man known to have codified the method called Double-Entry Book-keeping was Fra Luca Bartolomeo Pacioli, a Tuscan monk and a contemporary of Leonardo da Vinci. Pacioli included his account of accounts in a much larger mathematical work, *Suma de Arithmetica, Geometria Proportioni et Proportionalità*, which was printed in Venice in 1494" (17). In addition to proportionality, the double-entry system reflects a universe of stasis and imperturbable equilibrium. One arrives at the same offsetting results regardless of whether one records finances or perceived offenses.

For Henry Waugh the one trait shared by baseball and business is "Somebody to keep the books" (138). Record-keeping. Trying to impress Hettie Irden, a neighborhood B-girl he's met at Pete's Bar, he somewhat facetiously explains his role in the non-existent league: "I keep financial ledgers for each club, showing cash receipts and disbursements. . . . And a running journalization of the activity, posting of it all into permanent record books, and I help them with basic problems of burden distribution, remarshalling of assets, graphing fluctuations. Politics, too" (27). All fans of real baseball, of course, are statistics mongers. The game lends itself to record-keeping: each play can be reduced to an alpha-numeric shorthand; and statistics for various categories—home runs, sacrifices, strike-outs, and put-outs—can be compiled in the aggregate. Roy C. Caldwell, Jr. remarks (partly paraphrasing sportswriter Roger Angell) that baseball is "the most mathematical of sports. . . . The game provides a perfect, finished balance sheet: each accomplishment (credit) of the hitter represents a failure (debit) for the pitcher, and vice versa. Each individual match may be read afterwards from the composite of its statistics. From the complex network of real actions of the playing field, an absolute mathematical figure thus results. The scorecard and box score 'imitate' baseball, translate it into numerical representation" (163). If one were to go no further in this comparison than the observation that both the baseball box score and double-entry bookkeeping are zero-sum games, it would be difficult to imagine why Henry Waugh finds table-top baseball (which reduces the action to pure records-keeping) any more diverting or lively than his mundane chores at the accounting firm.[7] The point is that Henry doesn't just play and record games, he invents them.

Naturally, Henry is himself a Research Specialist in the Etiology of Homo Ludens, having "found it pleasant to muse about the origins" of his Universal Baseball Association:

> He'd always played a lot of games: baseball, basketball, different card games, war and finance games, horseracing, football, and so on, all on paper of course. Once, he'd got involved in a tabletop war-games club, played by mail, with mutual defense pacts, munition sales, secret agents and even assassinations, but the inability of the other players to detach themselves from their narrow-minded historical preconceptions depressed Henry. Anything more complex than a normalized two-person zero-sum game was beyond them. (44)

In the beginning there were closed-form games. The many exegetes of J. Henry Waugh (Jehovah) have commented that he is given to world-making.[8] But zero-sum games are inherently locked into static patterns, their participants unable to shake the orbits of historical preconception that lead them inevitably to the same conclusions: Wellington defeats Napoleon at Waterloo again; the New York Yankees with their Murderers Row sweep the Pittsburgh Pirates in four games in the 1927 World Series. Zero-sum games are thus a closed system that can be said to decline entropically either to a still-point of inactivity or to an endlessly repeated pattern. Their limited appeal rests in a mechanistic order, the confidence on the part of their creator that they will behave in a predictable manner (despite the presence of chance elements). But Henry remains disappointed with any closed-form game that merely replicates his monotonous and isolated life. Instead, he "invented for them a variation on Monopoly, using twelve, sixteen, or twenty-four boards at once and an unlimited number of players, which opened up the possibility of wars run by industrial giants with investments on several boards at once, the buying off of whole governments, the emergence of international communications and utilities barons, strikes and rebellions by the slumdwellers between 'Go' and 'Jail,' revolutionary subversion and sabotage with sympathetic ties across the boards, the creation of international regulatory bodies by the established power cliques" (44). Although he fails to entice the denizens of "battle games," Henry's creation of "Intermonop" (the competition of international cartels) appeals by reason of its openness, multiplicity, evolving strategy, complexity, and competition of subversive and regulatory factions. Like many experimental authors, Henry "did manage, before leaving the club, to get a couple pieces on his 'Intermonop' game published in some of the club literature" (45), but finds few readers.

Play Ball

Henry Waugh is the inventor and proprietor of a world-game. Rather than "escape" the zero-sum game of bookkeeping for an equally static contest, he devises a game based on complexity, procedure, compounding strategy, and multiple variables. His Universal Baseball Association is an open dynamical system that emulates the fluctuations of the universe at-large. Gazing at the image of a pitcher frozen in mid-windup on a pinball game at Pete's Bar, Henry declares such a limited simulacrum a "simple-minded" game: "In spite of all the flashing lights, it was . . . a static game, utterly lacking the movement, grace, and complexity of real baseball. When he'd finally decided to settle on his own baseball game, Henry had spent the better part of two months just working with the problem of odds and equilibrium points in an effort to approximate that complexity" (19–20). In this regard Henry resembles the author, who confesses a fascination with "highly complex structure" ("Robert Coover" 145). Coover discusses the evolution of his *Universal Baseball Association* with Gado, during which time he was "searching out the structure that seemed to be hidden in it":

> Even though structure is not profoundly meaningful in itself, I love to use it. This has been the case ever since the earliest things I wrote when I made an arbitrary commitment to design. The reason is not that I have some notion of an underlying ideal order which fiction imitates, but a delight with the rich ironic possibilities that the use of structure affords. Any idea, even one which on the surface doesn't seem very interesting, fitted with a perfect structure, can blossom into something

> that even I did not suspect was there originally. Engaging in that process of discovery is the excitement of making fiction. (148–49)

The "perfect game" thrown by Damon Rutherford in the first chapter of the novel is a figure for the author's search for that rare perfect structure. But Coover dismisses credulity in an ideal order that promises only an unchanging state of Being. His fiction reveals richer meanings in the ironic possibilities of an arbitrary form. In his use of design as a generative constraint for his fiction, Coover joins the postmodern turn to such devices as found in Walter Abish's *Alphabetical Africa* (1974), John Barth's *LETTERS* (1979), Italo Calvino's *The Castle of Crossed Destinies* (1973), Harry Mathews's *The Journalist* (1994), Milorad Pavic's *Dictionary of the Khazars: A Lexicon Novel* (1989), Georges Perec's *Life, A User's Manual* (1978), and Gilbert Sorrentino's *Aberration of Starlight* (1980). Acknowledging the arbitrariness of his method, Coover determines finally that the "game is not intrinsically so important; *what matters is that it be generative and exciting for me while I'm creating*" ("Robert Coover" 145, Coover's emphasis).

Henry's desire to intensify the complexity of his table-top game leads him to adopt the 56 combinations (based on odds of 216) of three dice and an elaborate system of cascading charts, culminating in the Extraordinary Occurrences chart triggered by a special combination of the dice. Henry's measures are intended to expand the possibilities and enhance the generative capabilities of his game. He creates a system with a capacity for generating new patterns and provoking unexpected organizations, comparable to positive feedback loops in cybernetics or autocatalysis in chemical reactions.[9] Such an open system resists a fixed or repetitive pattern. In the debacle following Damon Rutherford's death-by-pitched-ball in chapter two, Henry realizes that "the circuit wasn't closed, his or any other: there were patterns, but they were shifting and ambiguous and you had a lot of room inside them" (143). Room, that is, to continue playing. Jackson Cope suggests a perhaps too stringent dichotomy when he argues that "game is the opposite of play. Game implies an 'end,' a victory sought as the result of obeyed formulae with all of the statistics that Henry leans upon, the prop's props. Play is endless because pointless, mimesis of or escape from the unpredictable openness of causality" (38). But throughout the novel Henry defies the business world's edicts of punctuality, servility, and the all-important accomplishment of tasks. He doesn't want his game world to "end" but to sustain itself in an "unpredictable openness." The zero-sum box score records a single competition with a definitive outcome. There are no ties in baseball, though any game can be extended indefinitely in "extra innings" in the pursuit of victory. Henry, playing alone, playing both sides of the game, is less concerned with the isolated victory or defeat than in extending the process of his league. Neither the ultimate achievement of Damon Rutherford's "perfect" game, nor the crushing line drive that kills rookie pitcher Jock Casey on the mound, stop play. The game only acquires a "new ordering, perspective, personal vision, the disclosure of pattern, because he'd discovered . . . that perfection wasn't a thing, a closed moment, a static fact, but *process*, yes, and the process was transformation" (212). The Universal Baseball Association is an open system of evolving patterns because its maker has encoded it for fluctuation; Henry devises charts "for deciding the ages of rookies when they came up . . . and for determining who, each year, must die" (20). The influx of energy in the form of remarkable or disappointing rookies generates new orders—in the line-ups, in the standings—and the efflux of veterans who have expended their talents create change by their absence.

Coover accounts for the plenitude of game-players in his fiction by claiming, "We live in a skeptical age in which games are increasingly important. When life has no ontological meaning, it becomes a kind of game itself. Thus it's a metaphor for a perception of the way the world works, and also something that almost everybody's doing—if not on the playing fields, then in politics or business or education" ("Interview with Robert Coover" 72). Coover points to the absence of any transcendent, absolute order that would inspire belief. Under the condition of such skepticism, all customs appear arbitrary, provisional, and subject to manipulation. It's important to recognize that Coover doesn't adopt the concept of game as an antidote to metaphysical insecurities but as a metaphorical acknowledgment of "the way the world works." These games—in business, fiction, sport, or politics—are not escapist fantasies that turn away from confrontation with existence but are an honest recognition of the emergent and dissipating patterning of the world. Larry McCaffery rightly points to the rejection of a firm and inevitable causality in Coover's fictions, and thus the opposition to "*any* concept of an *externally imposed* system of order" ("Robert Coover" 110, McCaffery's emphasis) such as might be asserted by history, politics, or religion. McCaffery argues that two "alternatives are evident: either man can adopt the despairing outlook that life is fundamentally and irrevocably absurd and chaotic . . . or he can consider the freedom of each moment as a sign that we can create *our own* system of order and meaning." He concludes that characters such as Henry Waugh are attracted to games because they afford "order, definite sets of rules to be followed, a series of signs that can be interpreted, non-capricious rewards and punishments, and a sense of stasis and repetition that seems somehow free from the demands of process" ("Robert Coover" 110). McCaffery's existentialist diagnosis of the motive toward games and rituals in Coover's fiction relates to similar concerns in the postwar fiction of Barth, Saul Bellow, John Hawkes, and Kurt Vonnegut. But some additional comment seems needed from the perspective of recent investigation into chaotics. From the recognition that there are no absolute and transcendent orders, it does not follow that the world must be irrevocably absurd and chaotic. An exclusive dichotomy between a disorderly world and the ordering capacity of the human mind does not hold. Some patterns emerge unbidden in experience, even if they're not absolute; and many human actions invoke disorder, even when it has been provided for. As an extension of my earlier argument, although Henry adopts arbitrary rules in his gaming, his invention is neither static nor an escape from "the demands of process." Metafiction has frequently been charged with the invention of solipsistic orders that bear no relation to the condition of the world; irrelevance is a damaging charge. But Coover has already proposed that Henry's game incorporates the randomness of existence and provides a metaphor for "a perception of how the world works." Henry's invention is neither mimetic of a journey to chaos nor is it an escapist fantasy of an insulated ordering. The UBA is a form of orderly disorder, or as Coover says in his interview with Gado, an attempt to "furnish better fictions with which we can re-form our notions of things" ("Robert Coover" 150).

The UBA in the Balance

Henry Waugh rejects the closed system of order and the mechanistic simplicity of "a normalized two-person zero-sum game" in favor of the open system and complexity of his baseball league. This decision represents his rejection of a deterministic, Newtonian universe. Jackson Cope points out that the name of Henry's supercilious boss at the accounting firm, Zifferblatt, means "clock dial" in German (37). Henry creates flexible and unpredictable games (several of which are stuffed in his desk or concealed by his ledger) that defy the Zifferblattian conception of a clockwork universe in which all actions and reactions can be minutely calculated and accounted for. Henry's Universal Baseball Association as a world-order begins with the "problem of odds and equilibrium points" (*Universal Baseball* 20) in an effort to enact (not merely to simulate) the complexity of an open system that defies the calculations of linear dynamics. Strehle shows, especially in her discussion of Coover's *The Public Burning* (1977), that his fiction actualizes the shift from causality to probability, from the clockwork mechanism of Newton's universe to the radical indeterminacy of Werner Heisenberg's uncertainty principle.[10] In Henry's universe the laws of statistical probability take precedence over those of causality. After all, it's "the records, the statistics" that pique Henry's actuarial imagination; "real baseball," the actual collision of a wooden bat on a rawhide-covered ball propelling it in an arc of a certain distance and at a certain velocity, "bored him" (45). The curve ball defies the laws of dynamics (it has been called an optical illusion), because the ideal and time-reversible laws of motion disregard the effects of the atmospheric drag, the raised stitching on the baseball, and the sweat on the pitcher's fingers, as the ball spins toward the batter. Though it counters causality with probability, Henry's universe is not given over entirely to randomness. He envisions a system of "peculiar balances between individual and team, offense and defense, strategy and luck, accident and pattern, power and intelligence" (45). Henry fashions a game he hopes will be self-sustaining through the maintenance of a delicate and complex set of equilibria. He notes that "American baseball, by luck, trial, and error, and since the famous playing rules council of 1889, had struck on an almost perfect balance between offense and defense" (19).

It is necessary to distinguish between equilibrium structures in closed, static systems and those in the open, dynamical systems that describe the nature of life.[11] Let's take the example of a lock on the old Erie Canal, through which a barge wishes to pass. The west compartment of the lock has a higher water level than the east compartment. The difference in water level represents a store of potential energy available to do work. When the partition separating the two compartments is opened, water moves in an exchange between the two compartments until a level common to both has been achieved. The system has moved to a state of equilibrium. It "changes from a less probable to a more probable state and its entropy increases" (Arnheim 25). The tensions between opposite forces have been reduced, and the system arrives at a point of stasis—during which the barge may pass on from Buffalo to Albany. Equilibrium in closed systems describes a condition of non-differentiation and maximum entropy. As Rudolf Arnheim points out, equilibrium "represents the simplest structure the system can assume under the given conditions. This amounts to saying that the maximum of entropy attainable through rearrangement is reached when the system is in the best possible order" (25). Such an order, described by uniformity and inaction, is a numbingly simple condition, incapable of further change except by the introduction of external forces. If such a force were to be introduced, a hydraulic pump for example, it is possible to reverse the process and return it to its initial condition.

The equilibrium that Henry Waugh wishes to attain in his Universal Baseball Association is a different matter, more closely related to the concept of homeostasis in open systems that exchange energy, matter, or information with their environment. Homeostasis in a biological system as limited as a single-celled organism, or a social system as complex as a postindustrial city, is an orderly state sustained through the dynamical balance of opposing forces. Such a balance must not be a rigid opposition but a flexible process, one that through continual adjustment achieves an optimal relation between input and output. Arnheim comments, "Far from representing a striving toward deadly dissolution [of the sort one finds in thermodynamic equilibrium], the tendency to homeostasis ha[s] come about in biological evolution as a means of preserving life. Instead of the stagnation created by a state of maximum entropy, the open system of the organism constitute[s] a steady stream of absorbed and expended energy" (47). Henry Waugh's concern for equilibria is pitched precisely in an effort to evade the decay of the system to a state of maximum entropy or inertia and to sustain the long-run viability of the league's competitive process. An individual action such as pitching must be an interdependent relation of "power and control" (*Universal Baseball* 33), force and constraint. Waugh views his system as the creation of a life form, and it is no coincidence that he describes his sexual liaison with Hettie the B-girl in an extended metaphor of pitching: "In and out. Pitch and catch. Great game" (33). Each contest or series of games depends upon the multiple and minute adjustments between offense and defense to sustain competitiveness. Unlike a mechanical closed system in which a gain on one side is dependent upon an equal and opposite loss on the other, the open system fluctuates in a compensatory model. The defense acts independently of the offense, but it must compensate for a bunting situation by tightening the infield. Strategy and luck are both at work as the batter squares and the third-baseman charges. Strategy reduces vulnerability or creates opportunity; luck (either in the form of accident, or as a function of the differential between intention and execution) prevents invulnerability and magnifies risk. Equilibrium functions at the level of individual actions or contests and also at the level of the entire system. The league of eight teams, absorbing rookies and disgorging veterans, shuffling within the standings on either side of a .500 record, must sustain an equipoised competition. Paul Maltby notes the prevalence of the rhetoric of game theory in *The Universal Baseball Association:* "game theory is concerned with the question of the development of strategies that enable 'players' (e.g., business competitors, military opponents, political parties), usually in conflict situations, to maximize 'payoffs' without upsetting the equilibrium of the system in which they are 'playing.' This demands the study of the reasoning and dynamics of interdependent decision-making where one player's strategy depends on the likely strategy of another player and vice versa" (94). Maltby points to the UBA's Chancellor McCaffree and his reference to "compound decision problems" (*Universal Baseball* 146) and "correlated strategies" (152). Although Coover parodies this rhetoric as the unpleasant business-end of professional sports, and Henry as sole proprietor of the UBA is turned off by the militarism of two-party games, the correlated strategies and interdependent factors that Henry enumerates sustain the vitality of the league. Any occurrence that drives the system far from equilibrium threatens its collapse.

Device with Dice

The most prominent of interdependent factors that Henry investigates in his game are "accident and pattern." Henry "plays himself some device with dice" (234), an apt phrase whose rhyming terms should not go unnoticed. The Universal Baseball Association is no more than a device, an arbitrary yet intentional assemblage of performance charts and rules that express structure and reveal pattern. The league is Henry's invention, and to that extent it is subject to his control. But Henry counterbalances that determinism through the introduction of chance, the throw of the dice. Randomness sets the device into motion. The dice defy the determinism of Henry's universe, but they also propel it toward new and sometimes irregular, even remarkable constellations. Henry admits, "Oh sure, he was free to throw away the dice, run the game by whim, but then what would be the point of it? Who would Damon Rutherford really be then? Nobody, an empty name, a play actor. Even though he'd set his own rules, his own limits, and though he could change them whenever he wished, nevertheless he and his players were committed to the turns of the mindless and unpredictable—one might even say, irresponsible—dice. That was how it was. He had to accept it, or quit the game altogether" (40). Henry turns against authorial fiat—that the situation evolves the way it does only because he wills it so—and against an autotelic world in which every occurrence is predetermined by the laws that are inherent in the system (as suggested by Laplace). He embraces a probabilistic world in which any single event is unpredictable though there may be emergent patterns in the aggregate of all actions. The admission of randomness in the Universal Baseball Association should not be confused with a kind of Brownian motion. Henry's three dice are only a part of "the mechanics of the drama" (47). The elaborate performance charts in which three kinds of pitchers, Rookies, Regulars, and Aces throw to three kinds of hitters, Rookies, Regulars, and Stars (181), along with the possible invocation of the Stress Chart and the Chart of Extraordinary Occurrences, and special strategy charts for certain baseball situations, all suggest that while Henry may be slave to the dice he is still master of the game. Although the outcome of any one game cannot be predicted, there is no action that has not already been stipulated by the system of charts. Erase the distinction of fair and foul balls, have players move in Brownian motion rather than around bases, destroy the finely balanced structure of the game, and each contest becomes just as pointless as if Henry were to select the outcome at will. Only the interdependent combination of dice and charts, randomness and determinism, accident and pattern, ultimately has any meaning for Henry.

The Law of Averages

The Universal Baseball Association opens in the seventh-inning stretch of a pitching duel between rookie Damon Rutherford of the Pioneers and ace Swanee Law of the Haymakers. The game is scoreless to this point, and for those not attuned to the finesse of the game, it appears as though nothing is happening. But the moment is rife with dynamic tension. As a pitcher Swanee Law represents the Era of Equilibrium. The description of Law emphasizes his power and control, throwing heat and maintaining a low Earned Run Average. As a seven-year veteran and Ace of the Haymaker pitching staff, Law was "one of the main reasons Rag Rooney's Rubes had finished no worse than third from Year L through Year LIV. Ninety-nine wins, sixty-one losses, fast ball that got faster every year, most consistent, most imperturbable" of Haymaker moundsmen (4–5). On the verge of a neat one-hundred victories, Law is the figure of consistency and reliability; in a system that seeks to sustain its activities through regulation, Law is the most imperturbable factor. As a power pitcher "who just reared back and hummed her in," he nevertheless possesses "phenomenal staying power." His personality—like that of many star athletes—exhibited perhaps "too much steam" (5). The thermodynamic metaphor suggests an extremely efficient engine. Entropy is the quantitative measure of the amount of thermal energy not available to do work; it is also a measure of the disorder or randomness of a system. Henry Waugh refers to his table-top game as "the work" (5). For Henry's work to continue, he relies on the stability, predictability, and low entropy of players like Law. In a closed thermodynamic system, equilibrium will describe a state of maximum entropy in which little or no energy is available to do work. Law's regulation (neither too hot, nor too cold), however, maintains the UBA as an open system in a state of "balance" such that the energy injected into the system (nearly) equals the energy dissipated. Jackson Cope suggests that another analogy for Law's role in the novel is statistical. Swanee Law is the "law of averages, the opposite of Damon Rutherford who breaks them" (44). Law embodies the statistical mean, the tendency of the system toward the probable and the expected despite the occasional and unusual fluctuation. Like many students of baseball, Law is obsessed with charting his statistical progress: "Law knew what he had going for himself: whenever sportswriters interviewed him, they were shown large charts he kept tacked to his wall, indicating his own game-by-game progress in comparison with that of the five men in history . . . he was challenging" for season total strikeouts (*Universal Baseball* 144–45). As an Ace and strikeout leader, Law is himself no mediocrity; his comparisons are based on the performances of the best in the game. His capacity for "power and control," his resistance to perturbations, streaks and slumps, and his careful charting (in imitation of his creator, Henry Waugh, who records statistics for all games) allow him to predict, within a margin of error, his performance for the season. Thus he has "an outside chance . . . to hit the magic mark of 300" strikeouts for the season (which he does). Such reliability in the affairs of men inspires the "newsboys, bored, troubled, or revolted by what was happening—or not happening—in the rest of the league" (144, 145). Henry especially appreciates the compiling of records and statistics since, although it was the "static part of the game," it was "activity all the same, and in some ways more intense than the ball games themselves, a concentrated meditative concern with history, development, and equilibria" (205). Law has the appeal of a Satchel Paige or Nolan Ryan, whose careers were blessed by both the prowess of a power pitcher and the longevity usually associated with junk pitchers. In both table-top and real baseball, such a sustained performance is reassuring for spectators and participants alike. The equilibrium of an open system can be maintained through regulation, through the historical comparison of performance, through the development of the skills that the system depends upon for its survival. But for all his success, Law is still an aging veteran, an "old eagle" (144) who loses his pitching duel with the rookie Damon Rutherford. And Henry Waugh, aging at a different rate but now coincidentally the same age as his league, recognizes only in retrospect, "You mean, things were sort of running down before . . . ? Yes, that was probably true: he'd already been slowly buckling under to a kind of long-run market vulnerability, the kind that had killed off complex games of his in the past" (136; Coover's ellipsis). The obsessive regulation typified by Law cannot sustain an open system at a steady state indefinitely. Waugh must admit the encroachment of disorder and a gradual increase in the entropy of the system.

Rutherford's Dæmon

The rookie phenomenon Damon Rutherford provides the infusion of energy that the Universal Baseball Association needs. Henry observes retrospectively that "it was the kid who'd brought new interest, new value, a sense of profit, to the game" (136). More than just another source of excitement for the proprietor and only fan of the league, Damon acts as a negentropic force introduced into a system that was "sort of running down before." The second law of thermodynamics stipulates that the entropy of a closed system tends toward a maximum; that although energy is conserved, it is subject to a gradual dissipation. In open systems, however, it is possible for pockets of negentropy to coalesce, in which energy accumulates rather than dissipates, and order becomes more pronounced rather than decaying into disorder. As a rookie, Damon can be regarded as the introduction of an external source of energy into the "slowly buckling" league. But Henry discharges veterans and brings up rookies in the UBA every season—why should Damon's appearance perceptibly reverse the gradual decline of the game? In his pitching duel with Swanee Law, Rutherford functions as a dæmon that counters the Second Law of Thermodynamics by concentrating rather than dispersing energy.[12] The dæmon increases the orderliness of the system, reversing its tendency to decay and disorder. Damon's debut performance in the novel is not only a victory over the Law of averages and entropic dissolution but a "perfect" game in which all 27 batters are retired in order. In the ninth inning of the game, with the tension riding high, Henry recognizes the rarity of such an occurrence: "Odds against him, of course. Had to remember that; be prepared for the lucky hit that really wouldn't be lucky at all, but merely in the course of things. Exceedingly rare, no-hitters; much more so, perfect games. How many in history? two, three. And a Rookie: no, it had never been done" (11). Henry realizes that the system tends towards the more probable distribution of base hits and put outs; that is, towards a balance of offense and defense. Damon's feat is an exceptionally improbable occurrence, the apparently spontaneous collocation of a rare pattern: no runs, no hits, no errors. Henry expresses the human tendency to revere, and to mystify such nearly unique uprisings of order: "think what a wonderful rare thing it is to do something, no matter how small a thing, with absolute unqualified utterly unsurpassable *perfection!*" (23; Coover's emphasis). He appreciates the statistical improbability of such arrangements, noting that "there were other things to do, the record book was, above all, a catalogue of possibilities" (23). And yet the most unlikely thing has occurred. In fact the calculated complexity of the Universal Baseball Association fully accommodates the Extraordinary Occurrence (as Henry's charts signify) and thus makes aperiodic departures from the most probable and simple forms of play possible if unlikely. The homeostatic equilibration of an open system may sustain the system over a prolonged duration (Henry has logged 56 seasons in his record books and has introduced a second generation of players), but sustaining the system in the most probable and perdurable order does not necessarily stimulate interest or suggest higher purpose (Arnheim 48). Damon's perfect game represents a negentropic infusion of energy into the system that Henry does not wish to "dissipate" (24).

Far from Equilibrium

Damon Rutherford's perfect game nudges the Universal Baseball Association away from the steady state in which it has rested for the past several seasons. The slow erosion of competitiveness and "long-run market vulnerability" that Henry observed are an unfortunate effect of the system at equilibrium. "Damon had been a wonderful league tonic. The whole process had been slowing down, the structure had lost its luster, there'd been rising complaints about meaninglessness and lack of league purpose" (104). Damon's feat infuses the game—and Henry's enthusiasm for its purpose—with a higher level of energy and an assertion of meaningful pattern. A one-time deviation from the more probable activity of the league should send the league into a near-equilibrium state, one that may retard its eroding vitality and yet would be unlikely to have major consequences in its structure or performance. But Damon's success is considered an "epochal event" (15) that defies the commonplace. Henry treats himself to a celebratory cognac at Pete's Bar, adopts the persona of Damon as he pitches his affections to Hettie in bed that night (the single deviation from his monasticism), and is too preoccupied by Damon's heroics to balance his books correctly during the foreshortened workday he endures under Zifferblatt's watchful eye. Not only does Damon's graceful accomplishment alter the routine of Henry's life, but it also changes certain routine patterns in the play of the game. Fluctuations away from equilibrium in both personal and ludic actions are not dampened but begin to intensify. Henry relinquishes his usual indifference for the players' performance: "Damon Rutherford meant more to him than any player should. It had happened before, and it had always caused problems" (38). At first Henry thinks to "baby" Damon and ensure his assumption of Ace status by season's end. But by the next day he cannot resist an encore performance by his hero: "*He wanted to see Damon Rutherford pitch again tonight!* It wasn't the recommended practice to start a pitcher after only one day of rest, but it wasn't against the rules" (63; Coover's emphasis). Not a violation of the natural order, then, but a deviation from the norm that becomes intensified by the system's sudden instability. The system fluctuating at near-equilibrium is now about to receive a jolt that sends it into a far-from-equilibrium condition where nonlinear relationships prevail.

In his study of nonequilibrium thermodynamics Ilya Prigogine applies the term "dissipative structures" to systems that are exchanging energy and matter (and thus information) with their environment (*Order Out of Chaos* 12). Closed systems adhere to the second law of thermodynamics, for which the "entropy, S, of an isolated system increases monotonically until it reaches its maximum value at thermodynamic equilibrium" (*End of Certainty* 60). But in systems that are not isolated, it is possible to distinguish two components in the entropy change, dS. The first, d_eS, represents the flow of entropy across the boundaries of the system; and the second, d_iS, the entropy produced by the system. The total change in entropy for nonequilibrium systems can be expressed as the sum of the external entropy flow and the internal entropy production, $dS = d_eS + d_iS$. The internal entropy production will always be positive. But when energy flows into an open system the external entropy flow will be negative. It is possible for a nonequilibrium system to evolve spontaneously to a state of increased complexity (*End of Certainty* 64). As it spins away from equilibrium, the Universal Baseball Association expresses the characteristics of such a "dissipative structure."[13]

If Law represents the "imperturbable" force of regulation and the tendency to equilibrium, Rutherford is the dæmon who provokes the onset of turbulence. His next contest against Knickerbocker rookie pitcher Jock Casey quickly propels the finely balanced system away from equilibrium. Henry has overcompensated for the fluctuation in the system. In the third inning, with Rutherford still pitching hitless ball, two consecutive rolls of triple ones on the dice occur. According to Henry's own device, this brings the Extraordinary Occurrences Chart into play. "This was the only chart Henry still hadn't memorized. For one thing, it didn't get used much, seldom more than once a season; for another, it was pretty complicated. Stars and Aces could lose their special ratings, unknowns could suddenly rise" (69). This is the chart of catastrophe—on which all manner of mayhem ("a brawl could break out, game-throwing scandals could be discovered" [70]) might ensue—or of the unpredictable success of an "unknown" (such as Bobby Thomson's ninth-inning home run off Ralph Branca that handed the 1951 pennant to the New York Giants over the Brooklyn Dodgers, the momentous "shot heard round the world" that serves as the dramatic setting for the opening chapter of Don DeLillo's *Underworld* [1997], originally titled "Pafko at the Wall" for the forlorn Dodger outfielder who helplessly watches the ball disappear into the crowd). Ironically, such catastrophes or unimagined victories are accommodated in the design of Henry's complicated game. With Casey pitching to Rutherford, a third consecutive roll of triple ones results in *Batter struck fatally by bean ball*. Henry must adhere to the rules of his own invention, or all is meaningless; rather than intervene, he lays his hero down when the dice are cast. He can't avoid the collision that occurs, "leaning over the dice, trying to stop, trying to back up, force like the clashing of tremendous gears shrieked in his mind" (73). Damon's death is an accident, a random effect of a series of collisions. It is just as improbable a failure as his perfect game was an unlikely success. But it is a violent perturbation that sends both Henry and the system he's created into a chaotic phase, far from equilibrium.

When the game finally resumes with Casey still on the mound for the Knicks, what had been a close pitchers' duel, a rivalry, turns into an 18–1 rout in favor of the "bad guys." Henry's finely balanced system has entered a state of instability. He languishes over the loss of his endearing rookie and the phase transition from steady state to chaos:

> What had happened the last four or five league years? Not much. And then Damon had come along to light things up again. And maybe that was it: Casey had put out the light and everybody was playing in the dark. An 18-to-1 ball game, they *must* be playing in the dark! He watched them down there, playing in the dark, running around, tripping over bases, there in the dark, wallowing around in heaps of paper, spilling off the table edge— (136; Coover's emphasis)

As a result of these compounding perturbations, Henry's world has moved from a near-equilibrium steady state to chaotic instability. Prigogine argues that in such systems far from equilibrium the possibility exists for a bifurcation in which two new stable solutions may emerge from the prior condition of instability. The system "chooses" one of the possible branches. There is a probabilistic element, because nothing justifies the preference of the system for any one solution. Prigogine reveals that "there is an irreducible random element; the macroscopic equation cannot predict the path the system will take. . . . We are faced with chance events very similar to the fall of dice" (*Order Out of Chaos* 162). An example of a simple bifurcation is the so-called "pitchfork bifurcation" (Figure 5). In the case of the Universal Baseball Association, one branch of this bifurcation is the continuance of the league in some "new ordering" (211); the other branch, in which the steady state drops to zero, is "Exit from competition" (135). Like the business whose books Henry keeps, that dissolution "was both his prospect and his problem. . . . Henry had a kitchen full of heroes and history, and after heavy investment, his corporate account had suddenly sunk to zero. Accretion of wasting assets. No flexibility" (135). Systemic collapse cannot be disregarded, but Henry grasps for the "solution" in which, through continued perturbation, a second steady state emerges: "the circuit wasn't closed, his or any other: there were patterns, but they were shifting and ambiguous" (143).

Henry's decision to play out the season allows for the possibility that the

Pitchfork Bifurcation

Concentration X is a function of the parameter λ, which measures the distance from equilibrium. At the bifurcation point, the thermodynamic branch becomes unstable, and the two new solutions b_1 and b_2 emerge.

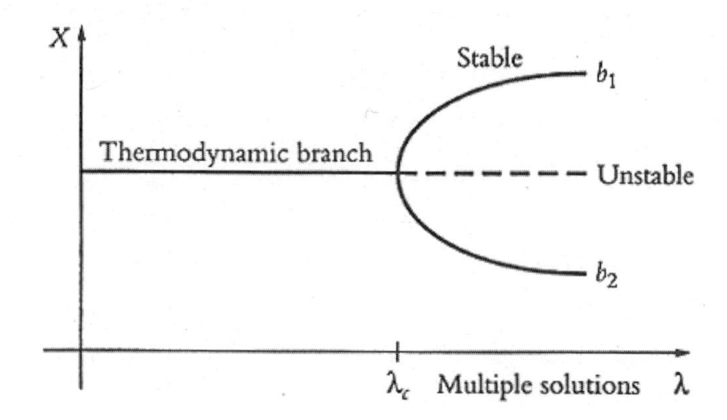

Figure 5. Pitchfork Bifurcation (reprinted by permission of The Free Press, a Division of Simon & Schuster, Inc., from *The End of Certainty: Time, Chaos, and the New Laws of Nature* by Ilya Prigogine)

second branch of the bifurcation will emerge, a new pattern from the midst of chaotic perturbations. This occurs not without further crises. The league Chancellor, Fennimore McCaffree (whose panoptic view of the game from his office makes him Henry's surrogate), observes that "his Association was undergoing a radical transformation, the kind sprung only from situations of crisis, extremity; his worries now were no longer merely political, but ones even of survival" (145). The stochastic leap in an unstable system to a new ordering—or to annihilation—is precipitated by the very extremity, the distance from equilibrium, of the chaotic phase. The Chancellor continues his analysis in the terms of game theory, though the relation to chaos theory is still apt: "What if, Woody, we have passed, without knowing it, from a situation of sequential compounding into one of basic and finite yes-or-no survival, causing a shift of what you might call the equilibrium point, such that the old strategies, like winning ball games, sensible and proper within the old stochastic or recursive sets, are, under the new circumstances, *insane!*" (148; Coover's emphasis). Despite his business-first approach, McCaffree perceives that the league has passed from a steady state (with its promise of extended process through compounding) to a phase of turbidity in which the now unstable system, acting chaotically, encounters a bifurcation point of either reordering or extinction. The leap to a new equilibrium point necessarily engages new strategies and new patterning.

Order Out of Chaos

Prigogine contends that self-organization—the increase in complexity observed in open systems—is "associated with distance from equilibrium" (*End of Certainty* 57). Stable systems will tend to dampen perturbations and fluctuations, followed by a return to a steady state. Equilibrium structures such as a crystal or stable pendulum are unchanging and therefore incapable of producing new order. Prigogine finds support in the philosopher Alfred North Whitehead's *Process and Reality* and in the physicist Erwin Schrödinger's *What Is Life?* for his argument that distance from equilibrium (or aperiodicity) and continual process are fundamental to the creativity of nature (*End of Certainty* 62). The evolution of a system, whether a living organism, a city, or an association of sports franchises, to a higher order of complexity requires the conditions of nonequilibrium. The final chapter of the *Universal Baseball Association* takes place on Damonsday CLVII, one-hundred seasons after the crucial bifurcation point in Year LVI of the league. The proprietor, J. Henry Waugh, is nowhere to be found in the post-apocalyptic Chapter 8, and the author refuses to comment on where Henry might have gone ("Interview with Robert Coover" 73). One theologically based speculation is that, having set his covenant with the death of Jock Casey, Henry recedes, never to intervene in the lives of his players again (Hansen 55). Much of the discussion among the next generation of players in Chapter 8 assumes an exegetical tone.[14] An argument to realism suggests that, having been fired from his job as an accountant, the unstable, insanely obsessed Henry becomes wholly absorbed in his enterprise to the exclusion of all other activities, rarely if ever emerging from his apartment. But we should consider the possibility that Henry is absent because he is no longer required by the system; the UBA in a state of nonequilibrium has self-organized to a higher order of complexity. There has been a transformation from the competitive game to the mythic reenactment of ritual.[15] The sacrifice of Damon "the dæmon" Rutherford and the retribution that falls upon Jock "J.C." Casey are now assumed as pre-scripted roles appointed to descendants of the original players. The Universal Baseball Association transcends the contingency of competition and the disparity between winners and losers in order to claim the greater predictability and reassurance of ceremony. Whether a player owes his allegiance to the deterministic necessity of the Damonites or the existential free will of the Caseyites, the league has evolved in an effort "to bring order to the chaos" (*Universal Baseball* 240) that marked the previous era. The rituals of Damonsday, unlike the rules and charts that governed the earlier games, are no longer the handiwork of Henry; they are devised by the league and the players as an attempt to understand the nature of their condition. From the nonequilibrium state that characterized the end of competition emerges the self-organization of the league as a complex system negotiating history, myth, and ritual. The evolution of the Universal Baseball Association need not be regarded as strictly governed by the physical laws of a system in thermodynamic nonequilibrium, but the fiction can be read as complementary in many respects to the phases of development in such systems. So for Coover the novel as design ultimately assumes a capacity for self-organization that extends beyond the control of its author. These complex systems—novels, games, and "dissipative structures" in the physical world—reveal a shared rather than isolated process of emergence.

7

The Excluded Middle

Complexity in Thomas Pynchon's *Gravity's Rainbow*

She had heard all about excluded middles; they were bad shit, to be avoided; and how had it ever happened here, with the chances once so good for diversity? For it was now like walking among matrices of a great digital computer, the zeroes and ones twinned above, hanging like balanced mobiles right and left, ahead, thick, maybe endless.

—Thomas Pynchon, *The Crying of Lot 49*

Science is a model of an "open system," in which a statement becomes relevant if it "generates ideas," that is, if it generates other statements and other game rules.

—Jean-François Lyotard, *The Postmodern Condition*

American literature . . . know[s] how to move between things, establish a logic of the *and*, overthrow ontology, do away with foundations, nullify endings and beginnings. They know how to practice pragmatics. The middle is by no means an average; on the contrary, it is where things pick up speed. *Between* things does not designate a localizable relation going from one thing to the other and back again, but a perpendicular direction, a transversal movement that sweeps one *and* the other away, a stream without beginning or end that undermines its banks and picks up speed in the middle.

—Gilles Deleuze and Félix Guattari, *A Thousand Plateaus*

If American literature expresses an affinity for the middle, then Tyrone Slothrop is the exemplary protagonist of Thomas Pynchon's *Gravity's Rainbow,* for he knows how to move between things—between the stars that chart his sexual conquests across the map of London, between the impacts of the V-2 rockets that fall on those very same quadrants, and between the restrictive control of his pursuers and the proliferating conspiracies of multinational forces in the postwar Zone. Slothrop's stuttered anaphora, "A-and," reverberating through the text articulates a grammar of infinite connectivity. The middle, free from the defining limits of form, represents a zone of complexity. Middles are expansive, seeking always to accrete and unconcerned to retract or expel what has already been encompassed. Middles are the zone of acceleration, in which vectors are multidirectional and not correlated, resisting the compression and regularization of linearity and targeting. Middles are comprised of a great many independent operators that interact with each other in as many ways. The complexity of *Gravity's Rainbow* should not be considered an aftereffect of the diverse materials that the novel subsumes; rather it is the methodology itself, to know a postwar world that is not running down but increasing in complexity. Since its publication in 1973, *Gravity's Rainbow* has been acclaimed a "difficult" novel. Were this so, then a dedicated reader's eventual mastery of its encyclopedic materials, perhaps with the aid of Steven Weisenburger's *Companion* to the novel's sources and contexts, or with reference to other critical commentary, should enable the reader to overcome that difficulty and arrive at a confident understanding of the text. Unlike highly allusive but formally constrained modernist texts, *Gravity's Rainbow* does not yield to a patient unpacking of its sources. Instead, it is a work of postmodern complexity. The reader's knowledge of its sources is not sufficient to explicate the complex array in style, conception, and narratology in which those materials are brought together. Such truly complex behavior is not reducible to its parts; to know *Gravity's Rainbow* is to know the complex patterning of the middle.

Monstrous Middles

Americans adore a humble beginning and a dignified end; for which, see the genre of the obituary. As for what passes in the middle, any number of controversies may ensue. Harold Bloom contends that *Gravity's Rainbow* is not "what is now called a 'text.' It is a novel, with a beginning, an end, and a monstrous conglomerate of middles" ("Introduction" 3). In attempting to preserve *Gravity's Rainbow* from "Derridean dissemination," Bloom appeals to the linear development of the novel form, however distended and distorted. (Is not one of the first characters we meet named "Teddy Bloat"?) Thus we have a text with "a" beginning and "an" end, but which is remarkable for its monstrously large midsection, an observation that would seem to confirm rather than dispel the indictment that *Gravity's Rainbow* is a rhizomatic text and not a linearly constructed novel. The book begins at the Chelsea maisonette of Pirate Prentice on December 18, 1944 in the preparation of a Banana

Breakfast; it ends in the Orpheus Theatre in Los Angeles, presided over by the night manager, Richard M. Zhlubb (a parody of Nixon) in 1970 in anticipation of the start of a film and the explosion of a missile.[1] In a highly episodic novel, these are set scenes that serve to launch or conclude the proceedings. They are brief, and their focal characters do not play essential roles in the narrative's midsection. Prentice, although a sympathetic protagonist who might easily serve as hero of many another novel of the war, largely disappears after the fifth episode of Part 1, "Beyond the Zero." But it is important that these two episodes, the beginning and ending of the novel, are locatable in place and time. Prentice has built a glass hothouse on the roof of his maisonette in which he cultivates bananas and other "pharmaceutical plants" (*Gravity's Rainbow* 5). One might compare Prentice's maisonette to Callisto's "hermetically sealed" apartment in Pynchon's early short story, "Entropy." Callisto creates a "tiny enclave of regularity in the city's chaos, alien to the vagaries of the weather, of national politics, of any civil disorder" (*Slow Learner* 83–84). He hopes to defer the entropic degradation of the isolated system he's established. In Prentice's rooftop hothouse, manure, "dead leaves off many decorative trees transplanted to the roof by later tenants, and the odd unstomachable meal thrown or vomited there by this or that sensitive epicurean—all got scumbled together, eventually, by the knives of the seasons, to an impasto, feet thick, of unbelievable black topsoil in which anything could grow" (*Gravity's Rainbow* 5). In the loam of such decay, and with the addition of energy from the sun (the hothouse is not a closed system), Prentice has created a pocket of negentropy in a violent and disorderly world, one that allows him to cultivate new life. As for the novel's end, it begins with Tyrone Slothrop's entropic "scattering" in the postwar Occupied Zone. He, more literally than other characters like Prentice who make a limited appearance in the novel, fragments into various alternate personae and finally disappears: "It's doubtful if he can ever be 'found' again, in the conventional sense of 'positively identified and detained'" (712). Zhlubb ushers the reader into a packed house to await "the pointed tip of the Rocket" of nuclear vaporization at ground Zero, as it "reaches its last unmeasurable gap above the roof of this old theatre, the last delta-t" (760). But some time before this Orphic dissolution the narrative itself has begun to disintegrate. The sixth episode of Part 4, "The Counterforce," fragments into twelve sections representing different genres and modes of discourse. The text has become, in the terms that Jean-François Lyotard reserves for the social collective, "a monster formed by the interweaving of various networks of heteromorphous classes of utterances (denotative, prescriptive, performative, technical, evaluative, etc.)" (65).

In truth the novel opens *in medias res*, "A screaming comes across the sky. It has happened before, but there is nothing to compare it to now" (3). The V-2 rocket is already en route between The Hague and London. The novel similarly closes with the nuclear warhead suspended, like Zeno's arrow, at the last possible delta-t above the Orpheus Theatre. And the audience is invited to join in the chorus of a preterite hymn, "Now everybody—" (760), an elliptic entreaty, trailing off. This refusal of absolute finitude suggests the circular form of James Joyce's *Finnegans Wake* whose final phrase interlocks with its initial phrase: "a long the . . . riverrun. . . . "[2] It would be more appropriate to say that the text opens and closes than to assert, along with the assumptions of exposition, development, and conclusion imposed by the genre of the novel, a Euclidean linearity that is implied by "beginning" and "ending." But it remains that the first and last episodes of the novel are locatable, "in the conventional sense of 'positively identified and detained.'" Bloom's compact appraisal of the form of *Gravity's Rainbow* is essentially correct. The novel is an infinite text confined in a finite space. It presents the paradox of an *open* (insoluble, indeterminate) *work* in a closed, determinate system of communication.[3] It offers a monstrously complex middle contained within a simple, or at least identifiable frame of reference. Somewhere between December 18, 1944 and September 14, 1945, one can locate the temporal cusp between the modernist era that concludes with World War II and the postmodern era incipient in the chaos of the Zone.[4]

Of beginnings and endings there is only one; of middles there are many. The neat singularity of any "incipit," the resolute finality of any conclusion, are reassuring in their naturalness. They are determinable. But middles are unpredictable and uncontainable. They are multiplicities capable of innumerable paths, uncharted. Bloom regards the middles of *Gravity's Rainbow* as "conglomerate," an indiscriminately adhesive mass. Although he considers the contemporary period to be "the Age of John Ashbery and Thomas Pynchon" ("Introduction" 1), Bloom's assessment of conglomerate middles does not award Pynchon points for high artistry. In fact there is more to these middles than adhesiveness. The heterogeneous cluster of middles in *Gravity's Rainbow* is born of Pynchon's essential confidence in the connectedness of creation, a position not far from the Symbolist poets' fascination with hidden correlations. The episodic structure of Pynchon's narrative—by which I mean the peripherally connected, quasi-independent status of many narratives, rather than a continuously unfolding of story—suggests a technique of multiplicity. One might apply the point of contention between the chemist and "cause-and-effect man" Franz Pökler and his wife Leni between direct causation and association to the narrative structure of the book: "'Not produce,' she tried, 'not cause. It all goes along together. Parallel, not series. Metaphor. Signs and symptoms. Mapping on to different coordinate systems'" (159). An "episode," a coming-in-beside, should not be regarded as a consequence or effect of the preceding action, but as a parallel mapping of discrete but related events.

Edward Mendelson's landmark study of *Gravity's Rainbow* as an encyclopedic narrative offers another explanation for "monstrous middles." Sheer size is a determinant factor, the incorporation of vast quantities and diverse sorts of information. Mendelson observes, "All encyclopedias metastasize the monstrousness of their own scale by including giants or gigantism" (164). The form of the encyclopedia assumes growth (accumulation of data; collation) from many points in many directions; that is, it effectively grows from the middle and not appreciably at the ends. No one reads an encyclopedia in sequence[5] from A, Aachen to Z, Zwingli.[6] The reader is encouraged to jump, or "meta-stasize," from place to place in the text. Naturally, the narrative structure of any encyclopedic text will be complicated by such shifts from one narrative space to another. But the distinctive function of cross-referencing in encyclopedic texts draws attention to connections between subjects and entries that would not be readily apparent in other forms of historical or argumentative narration. The presence of monsters, or gigantism, in the novel literalizes the insatiable absorption of the encyclopedia. For example, Pirate Prentice in his work as a "fantasist-surrogate" (12) encounters "*a giant Adenoid.* At least as big as St. Paul's, and growing hour by hour." The "lymphatic monster" has "a *master plan*" (ironically one of the few expressions of hegemonic intention in the book) to create a new elect and preterite as it absorbs its victims with a horrible-sounding "*sshhlop*" (14–15). The tentacular Octopus Grigori, "a gigantic, horror-movie devilfish" (51) presents yet another figure of the multidirectional, sinuous, unsegmented, and extensive grasp of the encyclopedic text. Both the adenoid and the octopus are versions of what Deleuze and Guattari call the "body-without-organs," particularly the *paranoid body*, whose "organs are continually under attack by outside forces, but are also restored by outside energies" (150).

Perhaps more important than the characteristic of exceptional size is the pathological nature of the monstrous. As Mendelson warns, "the book's ambition is essential to its design. No one could suppose that encyclopedic narratives are attractive or comfortable books. Like the giants whose histories they include, all encyclopedias are monstrous (as they are *monstra* in the oldest Latin sense—omens of dire change)" (165). Surely the premier site of pathology in the book is that which is ostensibly devoted to research and cure—"The White Visitation," housing a "catchall agency known as

PISCES—Psychological Intelligence Schemes for Expediting Surrender" (34). The extended description that is lavished on the appropriated mansion strongly suggests that deviation from the normal condition applies to both the investigators and their subjects.[7] A modest excerpt from the architectural catalogue seems warranted:

> The rooms are triangular, spherical, walled up into mazes. Portraits, studies in genetic curiosity, gape and smirk at you from every vantage. The W.C.s contain frescoes of Clive and his elephants stomping the French at Plassy, fountains that depict Salome with the head of John (water gushing out ears, nose, and mouth), floor mosaics in which are tessellated together different versions of Homo Monstrosus, an interesting preoccupation of the time—cyclops, humanoid giraffe, centaur repeated in all directions. Everywhere are archways, grottoes, plaster floral arrangements, walls hung in threadbare velvet or brocade. Balconies give out at unlikely places, overhung with gargoyles whose fangs have fetched not a few newcomers nasty cuts on the head. Even in the worst rains, the monsters only just manage to drool—the rainpipes feeding them are centuries out of repair, running crazed over slates and beneath eaves, past cracked pilasters, dangling Cupids, terracotta facing on every floor, along with belvederes, rusticated joints, pseudo-Italian columns, looming minarets, leaning crooked chimneys—from a distance no two observers, no matter how close they stand, see quite the same building in that orgy of self-expression, added to by each succeeding owner, until the present War's requisitioning. (82–83)

The narrator's description of "The White Visitation" as "erratic" and a "classic 'folly'" indicates a degree of repulsion for this monumental assemblage. And yet Pynchon is equally indulgent in his verbose articulation of the incongruously decorated interiors—molded plaster ceilings depicting a Methodist "peaceable kingdom," an Arabian harem, and a pig wallow (representing one of the author's eccentricities)—as well as an infinite regression of topiary in the English landscaping. The degree of elaboration here and throughout the book confirm Tom LeClair's thesis that *Gravity's Rainbow* illustrates the art of "excess" in American fiction (*Art of Excess* 43). More than an architectural critique, the figure of the White Visitation is a fractal iteration of the design of the whole book in its part. That is, one can identify important design features of *Gravity's Rainbow* embedded at a reduced scale in the *figura* of the mansion. The narrator recognizes an Escher-like pattern in the floor mosaic: the "tessellated" figures of the grotesques—made of small squares of stone, or tesserae—are interlocking and "repeated in all directions." These patterns are resistant to Euclidean geometry; there is no effective means for measurement or location within such infinitely repeated patterns. Several of M. C. Escher's drawings, such as *Day and Night*, *Sky and Water*, or *Mosaic II*, are fashioned from the endless tiling of images. The space they describe has no edge, but is comprised entirely of middles. Inevitably these expansive middles become monstrous, as they are here devoted to representations of "Homo Monstrosus." Thus the director of the White Visitation, Mr. Edward W. A. "Ned" Pointsman, F.R.C.S. regards the subject under study, Tyrone Slothrop, as "physiologically, historically, a monster. *We must never lose control*. The thought of him lost in the world of men, after the war, fills me with a deep dread" (144), a situation brought to pass by Slothrop's fragmentation (or tessellation) in the Zone. Slothrop, the architectural space of the White Visitation, and the space of *Gravity's Rainbow*, are similarly pathological, deviants from the normal condition.

Whether represented as a mosaic, labyrinth, lattice, or moiré, the middle space of *Gravity's Rainbow* is hideously complex, and more than a few readers have found themselves "walled up into mazes." Just as no two observers, no matter how close they stand, see the detail-encrusted White Visitation from quite the same perspective, so no two readers of the book arrive at quite the same interpretation. Weisenburger notes here a "visual example of indeterminacy resulting from parallax views" (56). The architecture of the White Visitation suggests a comparison between the period styles of the Baroque and the postmodern. Baroque art and architecture values the ornate and the complex over the austere and simplified. Postmodern architecture reacts to the suppression of ornament in modernism and the severe linear geometry of the Bauhaus. In the White Visitation we find the Baroque preference for elaborate ornamentation in "grottoes, plaster floral arrangements, walls hung in threadbare velvet or brocade," and for the balance of disparate parts in its "gargoyles . . . pseudo-Italian columns, looming minarets." In its eclecticism, postmodern architecture likewise adopts an assemblage of disparate styles. Although a conglomeration would subsume the distinctive qualities of its materials, the postmodern assemblage retains the stylistic identity of its component parts as a hybrid form (Jencks, *What Is Post-Modernism?* 50). It follows that Pynchon's *Gravity's Rainbow* is an edifice capable of appropriating various styles, subjects, and genres in a complex manner that relates these elements without conflating their distinguishing characteristics.

Further analogies can be drawn between the "monstrous middles" of *Gravity's Rainbow* and the peculiar class of geometric forms known as fractals, a term coined by the mathematician Benoit Mandelbrot. The most familiar

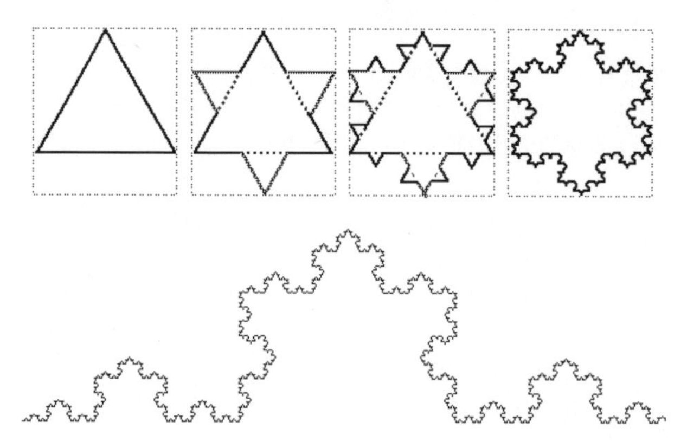

Figure 6. The Koch Curve (based on Benoit Mandelbrot)

of these forms, which is said to resemble a coastline or snowflake, is the Koch curve, first described by the Swedish mathematician Helge von Koch in 1904 (Figure 6). One begins with an equilateral triangle. At the middle of each segment, one adds a triangle that is one-third the size of the original. Repeat the process *ad infinitum*. The Koch curve grows from the middle. The polyp-like or crystal-flake shape becomes more and more finely detailed. Other fractal forms such as the Cantor set or Serpinski carpet are constructed by removing the middle sections of their initial line or square, respectively. James Gleick, in his treatment of Mandelbrot's fractal geometry in *Chaos*, describes the absorbing curiosity of the Koch curve. "Each transformation adds a little area to the inside of the curve, but the total area remains finite, not much bigger than the original triangle." One could create wallet-size Koch curves, beginning with a small enough triangle. "Yet," as Gleick observes, "the curve itself is infinitely long, as long as a Euclidean straight line extending to the edges of an unbounded universe. Just as the first transformation replaces a one-foot segment with four four-inch segments, every transformation multiplies the total length by four-thirds. This paradoxical result, infinite length in a finite space, disturbed many of the turn-of-the-century mathematicians who thought about it" (99–100). Although paradox may be upsetting to the objective and analytically minded mathematics community, Pynchon's fictions sustain many seemingly contradictory assertions. Like a Koch curve, *Gravity's Rainbow* constructs an infinite text within a finite space. Any reader who has turned from the book to other "source" texts knows that Pynchon exploits reference to other fields and knowledge-bases such as science, philosophy, and history.[8] Despite temporal limits on the present action of the novel, the narration frequently resorts to analepsis and prolepsis, making any moment in the text potentially expandable without limit. The use of symbology from various cultures—Teutonic, Herero, Norse—allows many embedded details and actions to assume deeply meaningful resonance. These literary techniques have been thoroughly expounded upon by critics of *Gravity's Rainbow* since its publication, and such methods are by no means unique to the work of Thomas Pynchon. In fact the use of reference, temporal shifts, and symbology represents an extension of the modernist practice of the novel of subjective consciousness that places Pynchon in the company of Henry James, Marcel Proust, Joyce, Thomas Mann, and William Faulkner. Yet Pynchon's infinite middles are largely unparalleled in postmodern practice. Gleick's assessment of the wonder expressed by mathematicians toward fractal forms has relevance to the critical response to *Gravity's Rainbow:* "The Koch curve was monstrous, disrespectful to all reasonable intuition about shapes and—it almost went without saying—pathologically unlike anything to be found in nature" (100).

The paradox—and the pathology—of a shape that describes an infinite perimeter within a finite area disturbs the conventional understanding of dimensionality. In Euclidean geometry, a point has zero dimensions; a line has one dimension (length); a plane has two dimensions (area); and space has three dimensions (volume). But Mandelbrot found that such measurement was only sufficient to describe regular forms. Just as Pynchon's book moves "Beyond the Zero," Mandelbrot found it necessary to move beyond integral dimensions of 0,1,2, and 3 to a fractional dimensionality in order to describe the irregularity of shapes in nature. Gleick comments, "Fractional dimension becomes a way of measuring qualities that otherwise have no clear definition: the degree of roughness or brokenness or irregularity in an object" (98). Whereas a "simple, Euclidean, one-dimensional line fills no space at all," the Koch curve, "with infinite length crowding into a finite area, does fill space. It is more than a line, yet less than a plane. It is greater than one-dimensional, yet less than a two-dimensional form. . . . Mandelbrot could characterize the fractional dimension precisely. For the Koch curve, the infinitely extended multiplication by four-thirds gives a dimension of 1.2618" (100–102). The more familiar curve in *Gravity's Rainbow* is the parabola of the V-2 rocket's trajectory which is a finite plane curve. But as Katje Borgesius remarks to Slothrop, "between the two points, in the five minutes, *it* lives an entire life. You haven't even learned the data on our side of the flight profile, the visible or trackable. Beyond them there's so much more" (209). Leibniz's calculus breaks down the arc into "slices of time growing thinner and thinner" (159).[9] But ideal parabolas subject to the analytic reduction of calculus will not represent the complexity, the irregularity found in nature. If the artillery could plot such regular forms without concern for the vagaries of wind speed and weather, trajectory and Brennschluss, the Nazis would soon have succeeded in obliterating every target in London. Rather than a smooth parabola, *Gravity's Rainbow* describes "the whole infinite self-embedding of complexity" (Gleick 100). Because Mandelbrot had not coined the term "fractal" until 1975, two years after the publication of *Gravity's Rainbow,* the book as a whole is a prolepsis of Pynchon's attraction to fractal dimensionality. That interest is demonstrable in *Vineland* (1990), in which Pynchon uses the term to describe complex phenomena such as the drifting aroma of breakfasts "sending out branching invisible fractals of smell" (323); and DL (Darryl Louise) Chastain's "light-bearing hair" is described, "against the simplicity out the window," as "a fractal halo of complications that might go on forever" (381). The infinite branching of fractal forms represents a complexity produced only in life. In *Mason & Dixon* (1997), one finds fractal symmetry in that staple of life, the British Loaf of bread. Charles Mason, Sr. "believes that bread is alive . . . the small cavities within exhibiting a strange complexity, their pale Walls, to appearance smooth, proving, upon magnification, to be made up of even smaller bubbles, and, one may presume, so forth, down to the Limits of the Invisible" (204). In all three novels Pynchon suggests that one encounters complexity on all scales in nature, both in the organization of the macrostructure and in the details of the microstructure. As with fractal forms, one finds a regular irregularity at every level of inquiry; and like the nourishing British Loaf of bread, *Gravity's Rainbow* exhibits complexity at every degree of magnification.

Excluded Middles

At the root of complexity, in *complector, complexus,* is the sense of to enfold, to braid, and to embrace. This gathering of the disparate into a compound entity does not eliminate the distinguishing characteristics of the components. Rather complexity represents the refusal to choose between two or more entities or possible states. Simplicity, by contrast, requires that a single entity be pure in its distinction; it is the insistence on choice. In a phrase very near the beginning of *Gravity's Rainbow* that has been taken by many critics as a metafictive commentary on the condition or texture of the book itself, Pynchon acknowledges the essence of the complex: "this is not a disentanglement from, but a progressive *knotting into*" (3). In disentanglement one detects the mind's penchant for the orderly, the linear, and the segmented.[10] But Pynchon's fictions (including Herbert Stencil's pursuit of the eponymous character in *V.*, and Oedipa Maas's inquiries after the Trystero system in *The Crying of Lot 49*) do not attempt to unravel plots but become complicit in a process of intrication. As in the natural tendency of a fishing tackle or sewing box, there is a knotting of diverse lures and threads together. In such an entanglement, there is no effective distinction between surface and depth. Intrication results in a substance or body whose condition or texture is roughly the same throughout. If one were to penetrate to the "core" of such intrication, one would find a degree of complexity that was comparable to that of the "surface." And so such a *knotting into,* if it were possible to turn it inside out, would appear essentially the same. Western epistemology insists on the hierarchy of function in component parts. The rational, totalizing mind recoils from phenomena that cannot be so distinguished. And yet, complexity that expresses what Mandelbrot dubbed "regular irregularity" is by far the more pervasive experience.

The folding, gathering together, or *knotting into* of complexity resists an explicit choice between alternatives, including those that are antithetical or antagonistic. The salesman Wimpe, discussing "a massive program to explore the morphine molecule . . . to find something that can kill intense pain without causing addiction," remarks that "We seem to be up against a dilemma built into Nature, much like the Heisenberg situation. There is nearly complete parallelism between analgesia and addiction. The more pain it takes away, the more we desire it. It appears we can't have one property without the other, any more than a particle physicist can specify position without suffering an uncertainty as to the particle's velocity—" (348). The exclusionary decision-making process of Western science insists on the isolation of parameters. Measurement cannot proceed without the selection of either one property or another; but as Werner Heisenberg's uncertainty principle proposes, the more exactness the investigator employs in measuring one property, the more indeterminacy results in measuring the other property. Wimpe recognizes such holism as "a dilemma built into Nature" because the pervasiveness of organic complexity resists analysis or separation into parts.

In addition to the analytic reduction into component parts, Western thought processes exhibit a binary reflex. Situational logic presents paired alternatives. Archetypal myths, philosophical concepts, and political discourse are frequently cast in oppositional terms. Binary logic demands that a choice be made between two terms, since it would be illogical to hold two contradictory thoughts in mind simultaneously. (The rhetorical use of irony would be one method, and for that reason it is usually excluded by scientific discourse.) It is to be expected that an encyclopedic text such as *Gravity's Rainbow* would contain references to a significant number of dualisms, some that are familiar to Western culture and some that are evoked for the specific occasion. As many critics have observed, the list of symbolic, conceptual, and character pairings in Pynchon's work is rather extensive. If one begins with a notable pairing such as the Elect (the chosen) and the Preterite (the passed over), brought to the fore by Tyrone Slothrop's Calvinist heritage, one can trace a series of related terms: the One (apex, or Brennschluss) and the Zero (the target, or ground zero); determinism and randomness; ascent and descent; the Schwarzkommando's Rocket 00001 and the Blicero's "Schwarzgerät" Rocket 00000; Germans and Hereros; Tchitcherine and Enzian; and so forth. In each pairing there is an almost unavoidable relation of dominant and subordinate between the two terms, one term that is effectively to be preferred over the other. The moral dualism of Pythagoras that found "the good" in all things orderly and "the bad" in all things disorderly, reinforced by Platonic logocentrism in the primacy of speech before writing, held sway in Western metaphysics at least until the deconstructive method of Jacques Derrida (in *Of Grammatology*) showed that the precedence of one term before the other is illusory because the idea of origin is an illusion.[11] Binary logic has enforced the either/or determination. Only in the Yin/Yang of Buddhist philosophy is the both/and complementarity of opposites strongly considered.[12] Slothrop expresses the tendency to assert preference between dualities, in this case paranoia and anti-paranoia, when he surmises, "Either They have put him here for a reason, or he's just here. He isn't sure that he wouldn't, actually, rather have that *reason*" (434). Given the invidious choice between a deterministic existence ruled by forces outside his Control, and random dissipation or Sloth, he might prefer the former poison to the latter. Nevertheless, Slothrop *qua* Slothrop represents, as Susan Strehle puts it, the "'zero' of discontinuity" to Blicero's "'one' of serial causality," dispersal of the self to the monomania of Blicero's Schwarzgerät (57–58). Slothrop expresses the anxiety of choice that is endemic to binarism, but he is also subject to representation within the system of dualism. Stepping out of that system of opposition, into the ungovernable middle ground of the Zone, is the challenge that Pynchon sets for his character (one that he does not, as such, survive) and for the reader.

After acknowledging that *Gravity's Rainbow* is replete with binary pairings that may be traced by the ambitious critic in the nearly inexhaustible pleasure of connection, one realizes that Pynchon could not leave such a hoary system unexamined. The binary pairing of Pavlovian psychologist Ned Pointsman and statistician Roger Mexico explicitly calls into question the reliance on binary logic in the scientific community. An obvious irony arises in Pynchon's use of such oppositional characters (his detractors would argue that they are straw men in a conceptual debate) to challenge the hegemony of binary logic in Western science. But irony's suspension of incongruity has been the literary tool of first resort in the struggle with scientific rationalism, and Pynchon employs it, naturally, as the antithesis to the binary mode of thought.[13] Pynchon draws attention to the oppositional relation of his two characters by casting them in the form of thesis/antithesis, much as he refers to paranoia's opposite elsewhere as "anti-paranoia" (434):

> If ever the Antipointsman existed, Roger Mexico is the man. Not so much, the doctor admits, for the psychical research. The young statistician is devoted to number and to method, not table-rapping or wishful thinking. But in the domain of zero to one, not-something to something, Pointsman can only possess the zero and the one. He cannot, like Mexico, survive anyplace in between. Like his master I. P. Pavlov before him, he imagines the cortex of the brain as a mosaic of tiny on/off elements. Some are always in bright excitation, others darkly inhibited. The contours, bright and dark, keep changing. But each point is allowed only the two states: waking or sleep. One or zero. "Summation," "transition," "concentration," "reciprocal induction"—all Pavlovian brain-mechanics—assumes the presence of these bi-stable points. But to Mexico belongs the domain *between* zero and one—the middle Pointsman has excluded from his persuasion—the probabilities. A chance of 0.37 that, by the time he stops his count, a given square on his map will have suffered only one hit, 0.17 that it will suffer two. . . . (55; Pynchon's ellipsis)

As a protégé of Pavlov, Pointsman adopts a method of psychological conditioning that proceeds according to strict determinism. Like Franz Pökler, he is a "cause-and-effect man" (159) who studies stimulus and reaction as a direct relation in accordance with Pavlovian laws set forth in "The Book," *Lectures on Conditioned Reflexes*. Pointsman's confidence in the material causality of psychological states is suggested by the Pavlovian "law of summation," in which the combination of "a number of weak conditioned stimuli" will result in "their exact mathematical sum" (Weisenburger 41). He is on common ground with Mexico only in their mutual distaste for the metaphysical psychic research of other White Visitation staff. Opposed to the possibility of any ethereal aspect of the psyche, Pointsman, like other behaviorists, pursues a relentlessly materialist view of the human mind. As with other absolutist solutions to vexing problems, there could be "No exceptions" to the "stone determinacy of everything, of every soul" (86). Thus the Pavlovian ideal is "the true mechanical explanation . . . a pure physiological basis for the life of the psyche. No effect without cause, and a clear train of linkages" (89). The specter of totalizing control arises because, as Alan Friedman and Manfred Puetz suggest, "if the Pavlovian assumption is right and everything can be *explained* mechanically, it follows that most things and events eventually can be manipulated and controlled that way" (27–28).

Although it would be anachronistic in 1944 to describe the cerebral cortex as a computer (the first electronic digital computer, ENIAC, was not completed until 1946), Pointsman conceives of the brain as a "mosaic of tiny on/off elements" (55). The brain becomes an instantiation of the binary logic of the digital computer. Or the computer is the most refined, powerful instance of the brain's capacity for binary thought. Like a program or cache of information comprised of binary digits, the Pavlovian brain occupies only the domain of the zero and the one, the not-something or the something. The sought-after determination and predictability of behavioral conditioning "assumes the presence of these bi-stable points" of excitation or inhibition, bright or dark, awake or asleep. States of ambiguity, indeterminacy, or probability, "anyplace in between" must be excluded from consideration. Pointsman endorses the Pavlovian "idea of the opposite" (87). Each behavioral state has a single coordinate opposite. Multiple or uncoordinated alternatives are not possible. For each behavioral state x, there must be a state not-x. All conditioning occurs in the alternation between these two "bi-stable points." Pointsman surmises that the conditioning of the infant Slothrop's erections by Dr. Laszlo Jamf in 1920 was in accordance with the "idea of the opposite": "a hardon, that's either there, or it isn't. Binary, elegant. The job of observing it can even be done by a student" (84). The subject of Pointsman's research is to discover the conditioned stimulus that accounts for the erections that Slothrop experiences preceding the V-2 rocket strikes on London. The putative solution (for those who attend to the story) is the aromatic polymer Imipolex G, "the first plastic that is actually *erectile*" (699), that Jamf develops for IG Farben and which is used as an insulator for the V-2 rocket.[14]

Friedman and Puetz show that among the "opposing groups of characters" in the novel, while "Pointsman, the Pavlovian, stands for strict determinism and predictability of material as well as psychic events, Roger Mexico, the statistician, represents the approaches of randomness governed by the laws of probability" (27). Thus Pointsman, the character most closely identified with binary logic in the book, is himself subsumed within a dualism for which Mexico, the Antipointsman, represents the opposite term.[15] And Mexico recognizes in Pointsman "The Antimexico. 'Ideas of the opposite' themselves, but on what cortex, what winter hemisphere? What ruinous mosaic, facing outward into the Waste" (89). Pointsman aspires to a mechanical determinism so thoroughly complete that no event, action, or behavior could occur purely by accident—contributing to a paranoid theory of history. But Mexico plays with "these symbols of randomness" so that he "wrecks the elegant rooms of history, threatens the idea of cause and effect" (56), countering with an antiparanoid theory of the 1960s vintage "happening," or the less aestheticized "shit happens." Mexico's generation (for he is clearly a prolepsis of My Generation) is that of the first postmoderns, affirmatively asking the rhetorical questions, "Will Postwar be nothing but 'events,' newly created one moment to the next? No links? Is it the end of history?" (56). Mexico counters the Pavlovian "idea of the opposite" with the "Poisson distribution" that maps both Slothrop's sexual conquests and the rocket strikes throughout the quadrants of London.[16] The rigid binarism that requires for each term a single opposite finds its own alternative in statistical distribution. The Poisson equation for the rocket strikes shows that they "aren't clustering. Mean density is constant" (56). The location of each strike is independent from that of all others. A strike on a particular quadrant does not enhance or detract from the probability of any subsequent strike falling on the same square: "the odds remain the same as they always were" (56). Mexico makes his methodological opposition to the material causality of behavioral psychology explicit: "Bombs are not dogs. No link. No memory. No conditioning" (56). The Poisson distribution counters Pavlovian binarism because it comprehends multiplicities rather than dualities, and most importantly, it refutes the causal links between phenomena that behavioral conditioning strives so diligently to achieve.

It's clear from this passage—which is cast as an omniscient analepsis (Weisenburger 40) to a debate at the White Visitation between Mexico's statistical and Pointsman's deterministic theories about the correlation of rocket strikes and Slothrop's erections—that Pynchon has established a binary opposition between these two men, both as characters and in the scientific methodologies that they represent. Pointsman's need for predictability compels him to seek results in terms of mutual exclusion—either/or, on/off, yes/no dichotomies. As a *points* man, he "can only possess the zero and the one." Mexico's statistical method dwells in "the probabilities," rather than sureties. He cannot reassure anyone that they are safe from attack. To him "belongs the domain *between* zero and one" (55), the middle rather than the ends, the fractional ground. This binary opposition within *Gravity's Rainbow* functions no differently from any other dualism. That Mexico represents the favored term should be clear to even the novice reader of Pynchon. In addition to his free-spirited love affair with Jessica Swanlake and his antiauthoritarian behavior (when as a member of the Counterforce he urinates on the Firm's conference table [626–40]), the openness—if not quite liberality—of Mexico's probability theory correlates with Pynchon's epistemology. One cannot make inquiries of an indeterminate and complex world that can be answered in finite, yes or no terms. Roger Mexico's probabilities, like Pynchon's ambiguities, occupy the domain of the "excluded middle."

The Law of the Excluded Middle was formulated by Aristotle in the *Posterior Analytics.*[17] As a law of logic, if there are two propositions, one of which directly denies what the other affirms, then one is necessarily true. Either A is B or it is not-B. No third term is possible (*tertium non datur*). The proposition that the patient is in a state of excitement (A is B), directly denies the proposition that the patient is in a state of inhibition (A is not-B). The Law of the Excluded Middle prohibits indeterminate or undecidable propositions that are *both* true *and* false. The patient cannot be in a state that is somehow *both* excited *and* inhibited. Such contradictory propositions cannot be held simultaneously. Pointsman, held in the thrall of binary logic, asks Mexico, "Can't you . . . *tell*" whether a place is safe—or not—from attack. But Mexico must answer "No." to the request for contrary, yes/no sureties. His probability theory occupies the domain of the excluded middle, a "chance of 0.37 that, by the time he stops his count, a given square on his map will have suffered only one hit" (55).[18] We've already established that *Gravity's Rainbow* contains many dualisms for which a ready preference can be expressed between the opposing terms; that is, Mexico's probabilities and Pointsman's determinism, the Counterforce and the Firm, the mensch and the Rocket, Rossini's romanticism and Beethoven's classicism (440), represent a series of decidable opposites that do not violate the prohibitions of Aristotelian logic. But there are at least as many instances of undecidable opposites in the book that fail to adhere to the law of contraries; that is, their two propositions are not directly deniable of one another. A familiar instance of the undecidable proposition occurs in the episode of Part 1 in which Roger Mexico and Jessica Swanlake on a lovers' jaunt happen upon an Advent vesper service in a small roadside church Somewhere in Kent. Although there is much unresolved critical debate as to whether the episode's indirect discourse is focalized chiefly through the consciousness of Roger and/or Jessica, one is entitled to surmise that some of the passage represents Roger's random thoughts during the caroling service and so provides some insight into the statistician's mind.[19] The passage anthropomorphizes the war, culminating in an "infant prince" who is brought "serai gifts of tungsten, cordite, high-octane." The war-child, however, proves inscrutable to the attendant kings: "what possible greeting or entente will flow between the king and the infant prince? Is the baby smiling, or is it just gas? Which do you want it to be?" (131). There appear to be two alternative reactions to the appeasement of the war-child's appetite: either the positive benediction of a smile; or the negative grimace of indigestion. The narrator inquires as to the preference of the narratee (whose diegetic, or extra-diegetic presence, is indeterminate), suggesting that "you" might well "want" an exclusive choice that reinstates the binary logic. But the propositions "the baby is smiling" and "the baby's passing gas" are not directly deniable. The expression on the war-child's face occupies the domain of the excluded middle; the propositions are *both* true *and* false: it may be a smile *and* it may be "just gas." Because this passage immediately follows a description of "a long-time schiz" at the White Visitation who "believes that *he* is World War II" (131), whom Mexico would have encountered in his work with Psi Section, it stands to reason that the characterization of the war within the domain of the excluded middle fits Mexico's probabilistic sensibility. The War's desires are complex. The figure of the war is irreducible to a single, determinate expression:

> The War needs to divide this way [into specialties and routines], and to subdivide, though its propaganda will always stress unity, alliance, pulling together. The War does not appear to want a folk-consciousness, not even of the sort the Germans have engineered, ein Volk ein Führer —it wants a machine of many separate parts, not oneness, but a complexity. . . . Yet who can presume to say *what* the War wants, so vast and aloof is it . . . so *absentee*. Perhaps the War isn't even an awareness—not a life at all, really. There may only be some cruel, accidental resemblance to life. (130–31; Pynchon's ellipses)

The narrator distinguishes between the totalizing propaganda of the German ideologues that subsumes or expunges difference and the phenomenon of the War itself that is not subject to a unifying control. Even though it rains death and destruction, the War resembles life in its complexity. It is not a governable, cybernetic "oneness," but an ungovernable multiplicity, a "machine of many separate parts" that operate independently of one another.

The most prevalent figure of undecidable opposites in the book is the mandala. These ritualistic geometric designs are inscribed on many objects with quasi-religious and cosmic significance, not unlike the post horn as a symbol of the Trystero in *The Crying of Lot 49*. The god worshipped by the Hereros is described as "creator and destroyer, sun and darkness, all sets of opposites brought together" (100) in the figure of a mandala. This sign of opposites together violates the logocentric, Western philosophical tradition. The complementarity of terms in the mandala conflicts with the law of excluded middles because it permits contradictory propositions to be held simultaneously as truth. Naturally it's Pointsman who decries "just this sort of yang-yin rubbish" (88) in Mexico's statistical theory. Like Pavlov, he anguishes over "the *ultraparadoxical phase* which is the base of the weakening of the idea of the opposite," rendering patients "paranoid, maniac, schizoid, morally imbecile" (49). He fights to prevent "this transmarginal leap, this surrender. Where ideas of the opposite have come together, and lost their oppositeness" (50). In contrast, the Zone Hereros construct the Aggregat 4 rocket—itself a communal counterforce to the monomania of Blicero's S-Gerät—according to the principles of mandalic force and complementary arrangement. When Slothrop presents him with a mandala he's found, Andreas Orukambe ritualistically interprets its symbols:

> "Klar," touching each letter, "Entlüftung, these are the female letters. North letters. In our villages the women lived in huts on the northern half of the circle, the men on the south. The village itself was a mandala. Klar is fertilization and birth, Entlüftung is the breath, the soul. Zündung and Vorstufe are the male signs, the activites, fire and preparation or building. And in the center, here, Hauptstufe. It is the pen where we kept the sacred cattle. The souls of the ancestors. All the same here. Birth, soul, fire, building. Male and female, together.
>
> "The four fins of the Rocket made a cross, another mandala. Number one pointed the way it would fly. Two for pitch, three for yaw and roll, four for pitch. Each opposite pair of vanes worked together, and moved in opposite senses. Opposites together. You can see how we might feel it speak to us, even if we don't set one up on its fins and worship it. But it was waiting for us when we came north to Germany so long ago . . . even confused and uprooted as we were then, we *knew* that our destiny was tied up with its own." (563; Pynchon's ellipsis)

The non-Western Hereros find peace, harmony, life and understanding in the mutually sustained "opposites together" where Pointsman, Blicero, and the other deterministic henchmen of totalizing rationality find madness, irresolvable conflict, and death. Notably the Herero mandala circumscribes a middle that, far from being excluded, contains their most sacred entities. The mandala represents a complete "*depolarizing*" (50) of Western logic. A scintilla of depolarization in Western thought is found in Niels Bohr's principle of complementarity (1927)—known as the Copenhagen interpretation of quantum theory—which sees "irreconcilable and mutually exclusive concepts—both particle and wave—as necessary to understand subatomic reality, which behaves according to both opposite principles" (Strehle 13). Yet Bohr's paradox takes only the first step in breaking the attachment to binary dualities in Western science.[20]

The reclusive Thomas Pynchon has assumed such a strongly countercultural identity that the endlessly divisive faction, or a cabal of one, would have to be the inevitable status of any movement with which he were associated. He has been labeled "subversive" (Tabbi, *Postmodern Sublime* 105) and a "dissident" (Maltby, *Dissident Postmodernists*); and before Theodore Kaczynski was discovered in his one-room cabin in remote Montana, Pynchon was under suspicion—even by his most avid readers—as the anti-academic, technophobic Unabomber, in part because of this subliminal suggestion made by a lone pine in the north German woods to Slothrop (whose family was in the New England paper business): "Next time you come across a logging operation out here, find one of their tractors that isn't being guarded, and take its oil filter with you" (*Gravity's Rainbow* 553). So it is not difficult to sustain the slight contradiction that, while Pynchon treats Mexico's statistical probability as preferential to the mechanistic determinism of Pavlovian psychology, he has not whole-heartedly endorsed any scientific methodology or philosophical system in the book.[21] Even though the two characters represent an antithesis of methodologies, Pynchon resists the urge to seize upon the indeterminacy, diversity, and distribution of Mexico's probability theory as the solution to the Slothropian enigma and other of life's mysteries. Pointsman makes the appeal that "We both have Slothrop" (89), but in fact neither do. The staff of the White Visitation propose various theories for Slothrop's proclivity for predicting the rocket strikes on London according to the prior distribution of his sexual liaisons. Rollo Groast suggests precognition. The Freudian Edwin Treacle argues for psychokinesis (85). And Pointsman will even entertain telepathy (89). But Mexico considers the coincidence of "girl-stars" and "rocket-strike circles" on the map according to a Poisson distribution to be a "statistical oddity" (85). It would be damn unusual (but not impossible) for two random patterns to overlap so closely (the effect is one of many moirés in the book). To accept such coincidence "implies moving . . . beyond the zero—and into the other realm" (85). Although Pointsman fails to identify the "Mystery Stimulus" he is willing to entertain preternatural ability in his methodical survey of possible connections. Mexico spurns causal connection in his application of the mathematical power series for exceedingly rare occurrences. As a pure statistician, Friedman and Puetz observe, Mexico "attaches neither hope nor desire to his view of the universe: 'Never had a prophetic dream, never sent or got a telepathic message, never touched the Other World directly'" (29). Just as Pointsman cannot isolate the conditioned stimulus that would provide a physiological cause for Slothrop's predictive ability, so Mexico cannot appreciate the preternatural aspect that distinguishes the correlation of the girl-stars and rocket strikes from mere chance. As oppressive as the predictability of a determinist universe may be, the purely mathematical analysis of probability offers little comfort in its disregard of the mysteries. Many readers of *Gravity's Rainbow* assume that Pynchon sustains the many dualisms of the book. But he is finally too subversive, too much a dissident to do so. Pynchon deconstructs the binary propositions of both behavioral psychology and probability theory to account for the Slothropian correlation. Thus Slothrop occupies the domain of the excluded middle; he is an undecidable proposition that is *both* true *and* false. Such theoretical systems of thought can never completely account for the complexity of his actions; they can provide only a schematic, or a calculus of approximation.

Complex Middles

At the Zero degree "nothing is connected to anything, a condition that not many of us can bear for long" (434). In this phase of pure randomness, no definitive patterns can be established; repetition, redundancy, surety and stability are absent; the particles or agents are free to move in a totally unrestricted manner. The zero state of randomness corresponds to the psychological condition of "anti-paranoia." In Mexico's statistical analysis "Everyone's equal. Same chances of getting hit. Equal in the eyes of the rocket" (57). Such randomness and disorder deeply frightens Jessica, because there is no resemblance to life; unrestricted freedom. At degree One "everything is connected" into a highly structured pattern, a crystal lattice of repetitive form. Control, predictability, and determinism are complete; predetermination, no free will. Deep and inescapable paranoia is of special interest to Pointsman, who has scribbled "exclamation points and *how trues* all about the margins of Pavlov's . . . Chapter LV, "An Attempt at a Physiological Interpretation of Obsessions and Paranoia" (87). Like the caged dog with its lip pinned up and salivary glands tapped, freedom of motion is totally restricted; freedom of thought constricted.[22] This too is a form of death. Only in the phase or state between the One and the Zero do we see anything that resembles life.[23]

The most recent scientific model to address the domain of the excluded middle is complexity theory. Although the specific theses of this discipline have been formulated in the last ten years (and so postdate the composition of *Gravity's Rainbow* between 1966 and 1971), the essential questions of how order might arise out of disorder, how complexity and life might emerge from the general tendency of entropic decay and homogeneity as described by the second law of thermodynamics, are among those that Pynchon's work repeatedly asks. There are now several studies of Pynchon's fiction that address the role of thermodynamics and the tendency of the universe "to bring events to Absolute Zero" (*Gravity's Rainbow* 3), the heat death that remains the ultimate prophecy of physics. Essays on the subject of entropy in Pynchon have been focused on the early short story, "Entropy" (now collected in *Slow Learner*), and on *The Crying of Lot 49*, where the relationship between thermodynamic entropy, as a measure of the energy not available to do work, and entropy in communications systems, as a measure of the efficiency of a system in transmitting information, is explicitly encountered.[24] There is an extent to which *Gravity's Rainbow* frets over entropic decay that might be drawn as a linear decline. But its preeminent image describes an infinite parabolic arc that completes itself in a mandalic cycle: "It Begins Infinitely Below The Earth And Goes On Infinitely Back Into The Earth it's only the *peak* that we are allowed to see, the break up through the surface" (726). Both the rainbow and the rocket's path are parabolic sections of a torus intersected by the plane of the earth.[25] The book holds out the prospect of a reorganization in the open system of the Zone that can be associated with the concept of negentropy, that in certain pockets of an open system there can be increases in order, structure, and complexity at the expense of disorder elsewhere in the system. Discussions of *Gravity's Rainbow* have emphasized the role that negentropy plays in an open system.[26] The thesis of self-organization in non-equilibrium thermodynamics, advanced preeminently by Ilya Prigogine in *From Being to Becoming: Time and Complexity in the Physical Sciences* (1980) and (with Grégoire Nicolis) in *Self-Organization in Non-Equilibrium Systems* (1977), underpins the discussion of the evolutionary structures proposed by complexity theory.

In his account of what he calls an emerging science at the edge of order and chaos, M. Mitchell Waldrop states that the capacity for *spontaneous self-organization* is the defining characteristic of a complex system. Whether at the molecular, biological, environmental, or cosmological level, these systems have the capacity for emergent order. To begin with, a system is *complex* because "a great many independent agents are interacting with each other in a great many ways" (11). A complex system can be distinguished from one that is merely complicated, in which there are many agents arranged within a restricted paradigm; or one that is disorderly, in which many independent agents do not interact in any meaningful way. Binary systems inevitably result in the polarization of the agents involved; the middle ground is cleared in the all-consuming recruitment of the singular and independent into one cohesive position or another. As Michel Serres remarks, "Thinking by negation is not thinking. Dualism is *noisy* or quarrelsome, its only connection is with death" ("Dream" 233). Complexity represents a depolarization, and the occupation of the middle ground. Pynchon's selection of the postwar Zone in which to set the bulk of *Gravity's Rainbow* reflects the sudden collapse of massive alternatives—the Allied and the Axis powers—into (for a time, because not all complexities remain unresolved) an area of multiple self-concerned and less strictly aligned forces, unofficial reprisals and shifting allegiances.[27] The independence of the agents in a complex environment is sustained because they are not locked into an oppositional relationship that permits only two kinds of interaction: either positive, as a reinforcement of kindred particles; or negative, as conflict with unlike particles. Thus independent agents are free to interact with each other in many more ways. In the place of rigid duality, one finds a more flexible multiplicity in the middle ground.

Waldrop distinguishes the complex system from the random, or Brownian motion, of particles by pointing out that from the molecular level to the ecosystem "groups of agents seeking mutual accommodation and self-consistency somehow manage to transcend themselves, acquiring collective properties such as life, thought, and purpose that they might never have possessed individually" (11). Complex, self-organizing systems are *adaptive*, actively responding to changing conditions in the environment and shaping their own behavior in such a way as to maximize their chances for survival. Adaptive agents are attuned to feedback from their environment (Waldrop 180). Pynchon pays obeisance to the "Cybernetic Tradition" of Norbert Wiener (although it's a slight anachronism in the present action of *Gravity's Rainbow* [Weisenburger 126]), and the "feedback system" (*Gravity's Rainbow* 238). Certainly Slothrop's survival in the disorderly flux of the Zone (Part 3 of the book) requires a sharp sensitivity to the changing conditions and a ready willingness to adopt other identities, such as Ian Scuffling (war correspondent), Rocketman (cartoon hero), a Russian officer, and Plechazunga (pig hero). Slothrop's adaptability helps him to maintain his freedom.

Complex systems flourish in the middle zone between restrictive order and destructive disorder. All "complex systems have somehow acquired the ability to bring order and chaos into a special kind of balance. This balance point—often called *the edge of chaos*—is w[h]ere the components of a system never quite lock into place, and yet never quite dissolve into turbulence, either. The edge of chaos is where life has enough stability to sustain itself and enough creativity to deserve the name of life" (Waldrop 12). We should not imagine that this "edge" of chaos requires such precarious balance that few systems could be established there for long. Rather consider a "zone between stagnation and anarchy" that encompasses the Brazilian rain forests, the Wall Street

Stock Exchange, the post-Soviet eastern European republics, and the primordial soup of amino acids from which life first arose (Waldrop 9). The middle zone of complexity between a totalizable, sterile orderliness and a randomized, disassociated disorderliness encompasses such a sizable proportion of the phenomena that one recognizes as life, culture and environment that one marvels at the lack of attention that complexity as a state-of-becoming has received. The Zone of *Gravity's Rainbow* is one intermediary area in which adaptive self-organizing systems can flourish.

The description of complexity as a phase state found at "the edge of chaos" is attributable to Chris Langton of the Santa Fe Institute in 1986. Langton expounded upon the research of physicist Stephen Wolfram that showed that the behavior of cellular automata[28] modeled on computers fell into four universality classes (Waldrop 225). The first three classes correspond to the phase states of dynamical systems. In Class I the cells converge upon a "fixed point attractor." No matter what the initial mix of "live" and "dead" cells, the entire system is drawn to a uniform, monochromatic condition of stasis. The rules for Class II behavior produce groups of cells that oscillate between two states. Although more motile than Class I behavior, Class II rules describe a "periodic attractor" that falls into a repetitive movement, like IBM's Deep Blue attempting to play out a perpetual check. The third class produces behavior of the opposite extreme, motion so turbulent that the display appears to be "boiling" with activity with no stable structures whatsoever (226). Class III rules correspond to the "strange attractor" in which no pattern is ever repeated exactly and behavior is unpredictable—the condition of chaos. However, Wolfram determined a fourth class of rules that produced activity that was neither resiliently static nor chaotically indeterminate. Class IV rules produced "coherent structures that propagated, grew, split apart, and recombined in a wonderfully complex way" (226). Or, in the full text of a poem by Robert Creeley, "Like life like" (41). Langton's investigation of Class IV rules clarifies the relation of complexity to static structure and chaotic activity. By adjusting the parameter for the probability that any given cell would be "alive" in the next generation, which he called lambda (λ), Langton would be able to gradually adjust the degree of activity among the cellular automata. At a lambda value of 0.0, no cells would survive much beyond the initial conditions, and a static system of the Class I type would result. If the lambda value were assigned at 0.5, "then the grid would be boiling with activity, with half the cells alive on the average and half dead. Presumably, such a rule would be in the chaotic Class III" (Waldrop 227).

When Langton set the critical lambda value at 0.273—just past the midpoint of his parameters between order and chaos—he found a cluster of rules that described sustained, self-generating complex systems. He was thus able to locate Class IV rules at the transition point between those describing the stable and unstable cellular automata classes: I & II → "IV" → III. This sequence also suggested that complex adaptive systems were to be found at a transitional moment between the orderly behavior of the periodic attractor and the disorderly behavior of the strange attractor in dynamical systems: Order → "Complexity" → Chaos (Waldrop 228).[29] Langton at first proposed the analogy that the transition between these states was like the phase transition of matter, with the lambda parameter akin to temperature. At low values of lambda, the Class I and II rules would "correspond to a solid like ice, where the water molecules are rigidly locked into a crystal lattice. The Class III rule that you found at high values of lambda would correspond to a vapor like steam, where the molecules are flying around and slamming into each other in total chaos," and the Class IV rules would correspond to the liquid phase of water (228). But further investigation by Langton proved that the analogy to phase transitions of matter was not an exact one. For one thing, gasses and liquids turn out to comprise a single fluid phase of matter. The "first-order phase transition" between solid and fluid is precise and abrupt. In the prosaic example of an ice cube raised to a temperature past 32°F, "the change from ice to water happens all at once. Basically, what's going on is that the molecules are forced to make an either-or choice between order and chaos" (229). In first-order transitions, the molecules of water are forced into a binary choosing between the lattice structure of the solid and the tumultuous movement of the fluid (though their "choice" is determined by the temperature). Langton, however, learned of "second-order phase transitions" that are less common, but also less abrupt because "the molecules in such a system don't have to make that either-or choice. They combine chaos *and* order." In such transitions one finds a roiling sea of molecules in the fluid phase, but interspersed and likewise tumbling in this sea are "myriads of submicroscopic islands of orderly, latticework solid, with molecules constantly dissolving and recrystalizing around the edges" (229). Although the system is predominantly chaotic, there is a mixture or distribution of orderly structures within it. In his poetic lecture, "Dream," Michel Serres extols this realm of the mixture that is "so intimate that there is not one chaotic or distributional area that is not surrounded with systems. . . . Nor is there one area of system that is not surrounded with distributions" (231). On a larger scale he views this "reciprocal plunge, this bathing of islands of order or negentropy in a fractal sea of rumbling" as itself a distribution (232). The second-order phase transition consists of a mixture or distribution of orderly islands dotting a turbulent sea in perfect balance. Langton found "order and chaos intertwine in a complex, ever-changing dance of submicroscopic arms and fractal filaments. The largest ordered structures propagate their fingers across the material for arbitrarily long distances and last for an arbitrarily long time. And nothing ever really settles down" (Waldrop 230). These second-order phase transitions represent a middle condition *between* solid and fluid. In place of a system in which the molecules are compelled to accept *either* a crystalline form *or* a fluid motion, the phase transition sustains a parity of *both* orderly structure *and* disorderly activity. The simultaneous suspension of alternatives can be compared to the indeterminate or undecidable condition of the excluded middle: are these islands of order in a sea of disorder, or a wash of disorder over archipelagos of order? And within this undecidable distribution there is a constant process of exchange, a reciprocal "becoming" of the one and of the other. As the fractal arms of the latticework reach out to seize and crystallize some molecules, others are dissolving and flowing away in its wake. Such a condition of becoming undoes and overwhelms the alternate states of being and not-being.

The complex adaptive systems modeled by Chris Langton, Norman Packard and others are distinguished by their ability to store information and share it with their environment. Self-organization requires that the system actively impute the data needed to raise itself to a higher level of complexity. Like a biogenic "in-box," the system must be able to store and sort the information it receives. The adaptability of the system functions as the "out-box," conveying information and responding to changes in the environment. Langton extends his analogy between phase transitions and complexity to computation, claiming that "Life is based to an incredible degree on its ability to process information. . . . It stores information. It maps sensory information. And it makes some complex transformations on that information to produce action" (Waldrop 232). Among the classes of cellular automata, the first two would be sufficiently stable to store the information, but their highly redundant structure would be far too passive to reorganize the information into more complex forms or to propagate information in the environment. Conversely the turbulent Class III automata would be far too noisy, pulsing with a constant stream of data, but without any capacity for retaining information or assembling it into units of greater complexity. Once again we find that only the Class IV rules "provide enough stability to store information *and* enough fluidity to send signals over arbitrary distances—the two things that seem essential for computation" (232). One needn't insist, as slaves to an analogy, that the mildly schizophrenic Slothrop behaves like a computational automaton in the Zone; but the complexity of his behavior may be explained by showing the relevance of Class IV rules "at the edge of chaos." After his escape from the Casino Hermann Goering, Slothrop behaves in a highly adaptive and improvisatory manner. He first assumes the identity of an English war correspondent, Ian Scuffling. The assumption of this rôle is, of course, not an accident. As a reporter he is in pursuit of information, specifically that related to his conditioning by the inventor of Imipolex G, Laszlo Jamf. He is "scuffling" in a disorderly struggle to make sense of his predicament and to save his own life. His first "contact" is the Russian black-marketeer, Semyavin. Instead of dealing in contraband, Slothrop's after information. Semyavin, with a mixture of disgust and resignation, exclaims, "Is it any wonder the world's gone insane, with information come to be the only real medium of exchange?" (*Gravity's Rainbow* 258). The wartime black market traded in the debasement of desires—drugs, sex, luxury items—and in the covert movement of wealth and currency (as revelations of Swiss banking practices have only recently shown). But in the postwar Zone we encounter the initial reckoning of a cybernetic system of information exchange. The phase transition marks a shift from the production of munitions and hardware tonnage to the packaging of international relations, treaties and trials; from industrial production to military-industrial espionage; from the "total war" and "scorched earth" of uncompromising opposition to realignment, reparation, and reconstruction. Semyavin laments, "Someday it'll all be done by machine. Information machines. You are the wave of the future" (258). Perhaps Slothrop is after all an automaton in the Zone, pursuing, consuming and propagating information. In computational terms, Slothrop on-the-run through postwar Germany is like an "undecidable" algorithm that might halt unexpectedly or continue gathering inputs indefinitely; there's no way to tell in advance which condition will be the case. Aloft in a balloon between the Hartz Mountains and Berlin, he is "without coordinates." In the foggy Zone, "Binary decisions have lost meaning" (335). As Waldrop points out, the only way to find out what "undecidable" algorithms will do is to let them run. Like Slothrop, "they exist in the only dynamical regime where complexity, computation, and life itself are possible: the edge of chaos" (234).

Just as Mexico's probabilities and percentiles for V-2 rocket strikes occupy the middle domain between Pointsman's binary digits, Slothrop enters the occupied Zone of Germany during the week after V-E Day, May 8, 1945, to skitter through a region characterized by confusion and unpredictability.[30] Slothrop's adherence to the Paranoid Systems of History has already led him to suppose that the bitter antagonism of the Axis and the Allies has been mutually abetted by a conspiracy of multinational corporations, including Shell International Petroleum, in the pursuit of greater profit: "Who'd know better than an outfit like Shell, with no real country, no side in any war, no specific face or heritage: tapping instead out of that global stratum, most deeply laid, from which all the appearances of corporate ownership really spring?" (243). The ability to recognize latent connections leads Slothrop into the middle realm of wartime profiteering where corporate entities with no alliances beyond their own global network exploit the bi-stable demand for munitions, technology, and financial services. The exploitation of the interstices of the wartime economy by international cartels such as IG Farben constitutes one of the major plots in a book in which "*Everything* is some kind of plot" (603). In the postwar disorder of the Zone, the gray areas of collusion and intrigue widen substantially. Slothrop learns from Pig Bodine and the Counterforce that "the arrows are pointing all different ways." The "red-pointed fingervectors" are evidently no longer aligned in his direction, because "the Zone can sustain many other plots besides those polarized upon himself." On the paranoid side of his cycle, Slothrop finds increasing complexity in the Zone's proliferation of conspiracy, so much so that the "network of all plots may yet carry him to freedom" (603).

But the anti-paranoid side of the cycle, in which nothing is connected, may prove even more unnerving. While in Zürich Slothrop meets the leader of a group of Argentine anarchists, "*magical precipitates out of Europe's groaning, clouded alembic*," or distillates of an alchemical conjunction of opposites to produce gold (Weisenburger 138). Squalidozzi presents the incessant political intrigue and revolution in Argentina and the promise of anarchic liberation in the newly created Zone: "It is our national tragedy. We are obsessed with building labyrinths, where before there was open plain and sky. To draw ever more complex patterns on the blank sheet. We cannot abide that *openness:* it is terror to us. Look at Borges" (*Gravity's Rainbow* 264). Somewhat later, the nonpartisan narrator notes that Borges, who was not sympathetic to his country's penchant for right-wing dictatorships, created labyrinths of uncertainty (383) rather than of imprisonment. Labyrinths of uncertainty such as the Zone do indeed represent the creation of "ever more complex patterns." The "rationalized space" (Weisenburger 138) of Daedalus's labyrinth (for which there is only one solution) is merely complicated—difficult to solve—but not complex or irresolvable.[31] The exiled Argentinian is enamored with the Zone's temporary lack of constraints:

> "In ordinary times," he wants to explain, "the center always wins. Its power grows with time, and that can't be reversed, not by ordinary means. Decentralizing, back toward anarchism, needs extraordinary times . . . this War—this incredible War—just for the moment has wiped out the proliferation of little states that's prevailed in Germany for a thousand years. Wiped it clean. *Opened it.*"

"Sure. For how long?"

"It won't last. Of course not. But for a few months . . . perhaps there'll be peace by the autumn—*discúlpeme,* the spring, I still haven't got used to your hemisphere—for a moment of spring, perhaps. . . . "

"Yeah but—what're you gonna do, take over land and try to hold it? They'll run you right off, podner."

"No. Taking land is building more fences. We want to leave it open. We want it to grow, to change. In the openness of the German Zone, our hope is limitless." (264–65; Pynchon's ellipses)

Slothrop can hardly comprehend the depolarized thinking of Squalidozzi, though he agrees to act as a courier to Geneva for a fee (making his tale another version of The Courier's Tragedy, introduced in *Lot 49*). The sudden openness of the Zone is a notable example of what Deleuze and Guattari refer to as "deterritorialization," as the constricted space of treatied fiefdoms is released into a smooth space in which anything can happen.

Slothrop's first location in the Zone (Part III, episode 2) is Nordhausen Mittelwerke (i.e., the "Central Works"), the enormous underground complex that Pynchon imagines as designed in the shape of the letters SS (the elite division's emblem, but "also a double integral sign!" [300], the method used by Leibniz for calculating trajectories) where the Germans assembled the V-2 rockets. As an introduction to life in the Zone, the Mittelwerke episode resonates with a sophisticated overtone series of mythic, historical, and scientific symbols.[32] In the immediate aftermath of the war, Mittelwerke has become the contested territory of the occupying armies in their pursuit of technological superiority: "English SPOG [Special Projectiles Operations Group] have come and gone. Right now American Army Ordnance people are busy crating and shipping out parts and tools for a hundred A4s. A big hassle. 'Trying to get it all out before the Russians come to take over.' Interregnum" (295). The U.S. Army held Nordhausen from April 11 to June 1, 1945. The historical moment passes between the stability of Germany as a nationalist order, the destructive disorder of the last battles of the European theater as the Allies drove on Berlin, and Germany's division into East and West republics in 1949. As Nordhausen, which lies just east of the eventual border between the divided sectors, passes from the control of one bloc to another, a zone of comparative instability arises in which cultural resources, technological information, and political power are commuted and exchanged. In the midst of a geopolitical transition, these "middle works" serve as a proving ground for complexity.

Among the prizes captured by the American Army at Nordhausen were the rocket engineers themselves who were allowed to make the transition from Nazi to NASA rocketry on the strength of their technological knowledge unencumbered by political sympathies or culpability in wartime atrocities. Pynchon considers with barely subdued irony the change-in-possession of the "A4 souvenirs" (295) and personnel from one political order to another without a notable interruption in their ballistic mission. Thus Mittelwerke assumes the self-determined functions of a Raketen-Stadt, or Rocket-City, a technological order that (for good or for evil) exists in between the competing geopolitical orders. The narrator suggests that the superficial symmetries of the Raketen-Stadt are reassuring to a military bureaucracy that seeks control over the mission and its powerful product. But Slothrop's descent deep within the Stollen, or tunnels, reveals a more intricate and less governable state: "Strangely, these are not the symmetries we were programmed to expect, not the fins, the streamlined corners, pylons, or simple solid geometries of the official vision at all—*that's* for the ribbon clerks back on the Tour, in the numbered Stollen. No, this Rocket-City, so whitely lit against the calm dimness of space, is set up deliberately To Avoid Symmetry, Allow Complexity, Introduce Terror (from the Preamble to the Articles of Immachination)" (297). The Raketen-Stadt allows for complexity by avoiding conscription in any reductive or a-priori forms. Like the Brennschluss point at the peak of the rocket's parabola, "It is most likely an interface between one order of things and another" (302). Rather than isolating one order from another, Pynchon superimposes pattern over pattern to create a moiré. Because one finds complexity at the edge or interface of order and chaos, Pynchon situates Slothrop in the Zone and his book at the very edge of a postwar, postmodern Europe.

The double integration of the V-2 guidance system presides over the incipient disintegration of Slothrop as a fixed or stable entity. He arrives in Nordhausen in the hope of discovering further truths regarding the connection of his childhood conditioning by Laszlo Jamf and the rocket. But the dark tunnels of Mittelwerke (and the adjacent prison camp Dora) are more conducive to the dissolution of the self than to its clarification. The lost souls he imagines there are "unique to the Zone, they answer to the new Uncertainty. Ghosts used to be either likenesses of the dead or wraiths of the living. But here in the Zone categories have been blurred badly. The status of the name you miss, love, and search for now has grown ambiguous and remote, but this is even more than the bureaucracy of mass absence—some still alive, some have died, but many, many have forgotten which they are. Their likenesses will not serve. Down here are only wrappings left in the light, in the dark: images of the Uncertainty. . . . " (303; Pynchon's ellipsis).

The middle work of *Gravity's Rainbow* is written under "the Ellipse of Uncertainty." The Rockets are supposed to "disperse about the aiming point in a giant ellipse" (425). There is a scattering, to be sure, in which each new impact redefines the arc of the ellipse, and yet gradually reveals its foci of intention. The trajectory of each Rocket varies independently from all others, but in the aggregate the falling Rockets discover a more complex pattern. Pökler the engineer reassures himself that the "chances are astronomically against a perfect hit, of course, that is why one is safest at the center of the target area" (425). Pynchon refuses to situate himself safely in the fixity of an exact center, but finds in the foliated burst of an infinitely expansive middle the risky space of his novel. If one sees the Ellipse of Uncertainty as a figure for the form of *Gravity's Rainbow,* then one also recognizes the rhetorical use of ellipsis as Pynchon's signature style.[33] In a sudden rush of paranoia, Pökler imagines, "why should he have been picked, unless . . . somewhere in his brain now two foci sweep together and become one . . . zero ellipse . . . a single point . . . a live warhead, secretly loaded, special bunkers for everyone else . . . yes that's what he wants . . . all tolerances in the guidance cooperating toward a perfect shot, right on top of Pökler . . . " (425; Pynchon's ellipses). The ellipsis for rhetorical effect is yet another excluded middle; of unspoken connections and implications . . . of paranoia and suspense. For Pynchon the device does not signify the omission of an unnecessary phrase but the recognition of a necessary connection. The novel occupies the space between the absolute focus of ground Zero and a dispersal so thorough that only a few segments of a vast ellipse are discernible.

8

The Superabundance of Cyberspace

Postmodern Fiction in the Information Age

The age demanded an image
Of its accelerated grimace . . .
—Ezra Pound, "Hugh Selwyn Mauberley"

"I can scribble faster than this!" Disraeli complained. "And in a better hand, by far!"

"Yes," Mallory said patiently, "but you can't reload the tape; bit of scissors and glue, you can loop your punch-tape through and the machine spits out page after page, so long as you push the treadle. As many copies as you like."

"Charming," Disraeli said.

"And of course you can revise what you've written. Simple matter of clipping and pasting the tape."

"Professionals *never* revise," Disraeli said sourly. "And suppose I want to write something elegant and long-winded. Something such as . . . " Disraeli waved his smoldering pipe. "'There are tumults of the mind, when, like the great convulsions of Nature, all seems anarchy and returning chaos; yet often, in those moments of vast disturbance, as in the strife of Nature itself, some new principle of order, or some new impulse of conduct, develops itself, and controls, and regulates, and brings to an harmonious consequence, passions and elements which seem only to threaten despair and subversion.'"

—William Gibson and Bruce Sterling, *The Difference Engine*

The paradigm shift from print to digital culture should be acknowledged as a defining aspect of postmodernism. As World Wide Web access in American households reaches the market penetration of such essential appliances as the microwave oven or the video cassette recorder, more individuals will turn to their computers to retrieve information than to any other source. For those already wired like a slipped disk into the backbone of the Internet, each day brings a surfeit of data: a barrage of e-mail, discussion-list postings, and "push" technology news; a proliferation of web sites devoted to subjects unclassifiable by any search engine; and a host of home pages populated by the notable or the impercipient. This shift in media culture has many enthusiasts. J. David Bolter, a prominent hypertext theorist and co-author of the hypertext writing system *StorySpace,* dubs our present condition the "late age of print" (quoted in Joyce 22). His use of the qualifier "late" implies that print is now in "a period of decline," though one can also read this phrase as heralding the death knell of print culture. Robert Coover, whose short stories in *Pricksongs and Descants* (1969) pioneered combinatorial narrative, likewise announces "The End of Books" in a review of some of the first published—or released—hypertextual works.[1] I find it less apocalyptic and more effective to regard the "lateness" of the hour of print media after the manner of Fredric Jameson, who ascribes to "late capitalism" not the "ultimate senescence, breakdown, and death of the system" as a function of entropy, but rather a "sense that something has changed, that things are different, that we have gone through a transformation of the life world that is somehow decisive" (*Postmodernism* xxi).

The transition to an electronic age also has its Cassandra whose dire prophecy is fated to go unheard. In *The Gutenberg Elegies* Sven Birkerts laments that "we are living through a period of overlap" between a print order that is linear and static and an electronic order that is interconnected and dynamic (122). These modal changes in the transmission of knowledge are not without consequences. The "imperatives of syntax" on the printed page and in the bound book enforce a hypotactic order that demands the reader's attention. Yet in the graphical clutter, hyperlinks, and bulleted lists of the web page "impression and image take precedence over logic and concept, and detail and linear sequentiality are sacrificed" (122). Birkerts contends that "we are in the midst of an epoch-making transition; that the societal shift from print-based to electronic communications is as consequential for culture as was the shift instigated by Gutenberg's invention of movable type" (192).[2]

When worlds collide. The period of overlap between the two media cultures of print and hypertext results in a condition of instability in which competing values and practices coexist. The widespread use of "officing" and desktop publication software has increased rather than reduced the paper traffic in the contemporary business environment. A print magazine advertisement for International Paper states boldly that the coming of the digital age has not created a "paperless society," only greater and more specialized demands for paper. We're told, "people don't just want information at their fingertips. They want it on their fingertips. They want to be able to touch, fold and dog-ear; to fax, copy and refer to; scribble in the margins or post proudly on the refrigerator door. And, above all, they want to print out—quickly, flawlessly and in vibrant color." Yet this multinational corporation, with its considerable investment in "providing the 'Paperless Society' with all the paper it needs," cannot neglect to announce its presence in cyberspace by listing its web page address (www.ipaper.com) in the small print. To cope with the persistence of this dual rubric (in both senses of "a class or category; a title or heading, written in red") of print and digital culture, many popular products display signs of cross-purposing: the requirement of textual and graphical links in web browsers; books-on-tape for the harried commuter or the visually impaired; electronic Post-it® notes for video display terminals; and full-text CD-ROM editions of fiction and non-fiction with "view" and "print" options.[3] Most daily newspapers and nationally circulated magazines are now accessible both at the newsstand and on the Web (sometimes in an abbreviated format to protect circulation, but often with multimedia supplements that expand on the print edition). The transition between paradigms is always marked by the presence together of functions that are incommensurable.[4]

In his foundational study of the convergence of hypertext and critical theory, George P. Landow contends that previous "transitions from one dominant information medium to another have taken so long—millennia with writing and centuries with printing—that the surrounding cultures adapted gradually" to the new paradigm (*Hypertext 2.0* 288). But in the age of technology, he suggests, an accelerated transition between media should be anticipated. The history of the recording industry in the twentieth century would seem to reinforce his point: the compact disc swiftly vanquished the long-playing phonograph record; hand-held video cameras have driven Super-8 film from the home market; and the newest delivery system for at-home cinema, Digital Versatile Discs (DVD), has already made substantial inroads in the market previously dominated by the VHS-format VCR.[5] Landow concludes, "This acceleration of the dispersal of technological change suggests, therefore, that the transition from print to electronic hypertext, if it comes, will therefore take far less time than did earlier transitions" (*Hypertext 2.0* 288–89). Possibly so, but Landow passes over the competing conveniences of print technology that will no doubt extend—perhaps for quite some time—a complete transition to electronic hypertext, such as the relative ease of handling paper in various places and postures, the portability of the book, the greater resolution of print text over screen view, the ease of mark-up with printed documents, and the tactile pleasure of fine paper.

The acceleration of technological change may also dissuade the general reader from mastering any system of digital text delivery. Platform obsolescence punishes the early adopter. The introduction of the Rocket Book in 1998 makes a hand-held portable reader of digital texts available to the public at a price of $499. Downloading texts from a single source in Barnes and Noble booksellers—at a price not less than the lowest retail price of the printed book—still requires an investment in a personal computer. And it's immediately apparent that such a single-purpose electronic book has little more functionality than an inexpensive trade paperback. Thus on another occasion Landow appears to defer the surprising conversion: "if the coming of print technology in the age of Gutenberg offers a relevant pattern, one can expect that old and new information technologies will exist side by side for a long time to come. If that situation comes to pass, one can also expect that, like video and film, hypertext will have an impact on materials written for print presentation" (*Hyper/Text/Theory* 40).

In fact there exists a complex dynamics of incommensurability that arises in periods of technological overlap. In one model the new paradigm engages in a dialectical struggle that supplants the practices and forms of the old. But another model of "intermediality" suggests that the introduction of new media can instigate a "catalytic interaction and intertwining of media from different historical moments" (Tabbi and Wutz 10). In this model the incommensurability of print and digital media serves to incite creativity in—and thus disturb, but not eradicate—the older, established forms of the printed text. Modulations in the form and method of the novel—the concept of what a "novel" might consist of, how its material might be manipulated—are provoked by the introduction of the competing paradigm. While the introduction of photography did not supplant painting in the nineteenth century, the documentary detail of the photograph compelled the artist to examine the conventions of mimesis, challenged the genteel rules of subject matter, and led to the foregrounding of painterly material in Post-Impressionism. In the twentieth century television has not eliminated radio as an important communications medium. The once staple genre of radio drama has virtually disappeared; but a marriage of convenience to telecommunications has spawned the call-in talk show as the most popular radio format. The previously passive audience has now become the interactive supplier of radio's broadcast material. Hypermedia could serve as a catalyst for a period of invention in print fiction akin to that of Post-Impressionism. The disturbing presence of hypertext and electronic media will not make print fiction disappear; but its presence will affect the manner and matter of the novel, making the compelling fictions those that generate associative logic instead of causal sequence, parallel structures (Milorad Pavic's *The Inner Side of the Wind, or The Novel of Hero and Leander* in fiction, John Ashbery's "Litany" in poetry, or Jacques Derrida's *Glas* in critical theory) instead of serial. In hypertext and print fiction the reader's apprehension of textuality, or what Jerome McGann calls the "textual condition," displaces the conduit metaphor of communication; reflexivity of the narrative dispels the reader's absorption in the text-world.

This disturbing transition between print and digital communications systems can also be discussed in terms of chaotics. As postmodern fiction enters a "phase transition" in its medium, the attendant disorder may yet be generative of new structures, just as new forms of order may arise in a dynamical state that is far from equilibrium. The nonlinear communications networks that enable hypertextuality have drawn the attention of literary theorists and critics who find that nonlinear narrative—even in print texts—is rather more the rule than the exception (Landow, *Hypertext 2.0* 43). This discovery of the nonlinearity of textual space arrives at much the same cultural moment as the investigation of nonlinear dynamics in the physical world reveals that nonlinear systems are the rule of nature and linear systems the exception. The work of the novelists presented in *Design and Debris* provides evidence for such a theoretical correlation. The postmodern "disruptors" in my discussion, Kathy Acker, Don DeLillo, and Thomas Pynchon, are strongly averse to the predictability of linear systems. Novels such as *Empire of the Senseless, Libra*, and *Gravity's Rainbow* are already decentered communiqués that resemble hypertextuality in their proposition of multiplex connections and in their resistance to a single channel of causality. They are "dissipative structures," in the sense given by Ilya Prigogine, because their disorderliness underwrites the capacity for an emergent self-organization. The postmodern "proceduralists" that I consider, John Hawkes, John Barth, and Robert Coover, realize an affinity for the cybernetics of command and control. Novels such as *Travesty, The Last Voyage of Somebody the Sailor*, and *The Universal Baseball Association*, implement deterministic rules that produce complex results in a nonlinear system. These texts engage the reader in a dynamical feedback loop that the point-and-click technology of hyperlinks makes rather literal. Like Edward Lorenz's "strange attractor," these novels reveal pattern in unpredictability. It seems inevitable that from a medium in turbulence new forms in fiction would arise.

I find no reason to lament the decline of five-hundred years of print technology and the erosion of readerly behavior; nor do I confess to a rapturous conversion to the incipient dataforms and their promiscuous linking. Rather I regard this transitional phase between the print and electronic orders as an unprecedented opportunity to study fiction writing at the precise moment in which the medium by necessity exchanges established compositional methods and organizational structures for inventive ones.

Hypertext revolutionaries consign print culture to the slow burn of acid decay. The promising pioneers of hypertext fiction and theory such as Michael Joyce, Stuart Moulthrop, and Jane Yellowlees Douglas are composing almost exclusively for an electronic environment.[6] These "writers of light," however, have yet to match for sheer fictive power and inventive genius such authors as John Barth, Italo Calvino, Julio Cortázar, B. S. Johnson, Milorad Pavic, Georges Perec, or Christine Brooke-Rose—all of whom anticipate hypertextual linking and nonlinear narrative in their fiction. The innovative writers whose work defined American postmodern literature since the 1960s learned their trade before the advent of the personal computer. Although the protagonist of Don DeLillo's *Underworld* visits a web site devoted to miracles (the supernatural kind, not those associated with sports events),[7] the author himself has said that he cannot compose without the physical "clack"—the rhythm and sound—of a manual typewriter's keys against a sheet of paper.[8] John Barth amuses himself at the expense of those who only point and click. In his short story, "Click," he parodies "The Hypertextuality of Everyday Life" by speckling his text in the venerable *Atlantic Monthly* (one of the oldest continually published journals in America) with the iconic rendition of hypertext links; the story critiques hypertextuality rather than actually engaging in it. As one might expect from this master of metafiction, a self-reflexive portrait of the "author" arises when his character Irma "clicks on **scene** and sees what the Author/Narrator sees as he pens this: a (white adult male right) hand moving a (black Mont-Blanc Meisterstück 146 fountain) pen (left to right) across the (blue) lines of (three-ring loose-leaf) paper in a (battered old) binder on a (large wooden former grade-school) worktable all but covered with implements and detritus of the writer's trade" (82).[9] Barth, like DeLillo, comprehends the appeal and activity of digital culture but portrays the "author" as an old-fashioned scribe, bracketed in the methodology and material of print culture.[10] No amount of hotlinked wizardry, however, will keep the information age client-reader glued to the computer screen. Only when the imaginative verve—which has little to do with computer savvy—that our most important writers possess has been brought intrinsically to the hypertext environment will reading hypertext fiction on a CD-ROM seem both compelling and natural. The transition from print to electronic texts, especially for the novel, won't be complete until there has been at least one, dare I say "masterly," author for whom electronic composition is the preferred medium, or more pertinently, the only medium that writer has ever known. That author would exploit—or perhaps I should say orchestrate—all of the intrinsic capabilities of a linked, multimedia text, so that "reading" the work would make the use of a computer indispensable.

Yet still there are postmodern novelists such as William Gibson, Kathy Acker, Richard Powers, and DeLillo, though bound to the print order, whose work is provocatively enhanced by an engagement with the terms and conditions of the information age, who invoke the cascade of associative thought that characterizes the experience of the Internet, and who have collaborated on web pages with digital artists and publishers, or are themselves the subjects of scholarly pages and popular discussion lists. These writers refuse to consign the novel to senescence. Nor does their art cower before the profusion of electronic media. Each engages the communicative act in the most present and prospective of forms.

It's Not O.K. to Be a Luddite

The postmodern world of cyberspace has departed—with mercurial ease—from the page-turning textual strategy of linear order for which Birkerts pines. To begin with, there is no center to the network of educational, military, and commercial mainframe computers that we call the Internet. The shortest distance for the transfer of files or exchange of electronic mail may not be a straight line; the complete header or "spool" for a piece of e-mail correspondence sent to a colleague in Hong Kong describes a wandering route from node to node before it reaches its destination. One's virtual presence on the Internet makes irrelevant the concepts of center and periphery; the center is wherever one's own interest and attention lies. The World Wide Web—as an ever-expanding mesh of nodes and cable—can be considered as an electronic manifestation of the rhizome, introduced by Gilles Deleuze and Félix Guattari in *A Thousand Plateaus*. They distinguish between an arborescent system with root and branches, and the rhizomatic system that spreads along multiple lines and connections. There is a family resemblance among webs, meshes, and weedy plants like kudzu. In the root directory and command tree of Microsoft's Disk Operating System one recognizes an arborescent system. Such systems, Deleuze and Guattari tell us, "are hierarchical systems with centers of signifiance and subjectification, central automata like organized memories. In the corresponding models, an element only receives information from a higher unit, and only receives a subjective affection along preestablished paths. This is evident in current problems in information science and computer science, which still cling to the oldest modes of thought in that they grant all power to a memory or central organ" (16). Premonitory of Microsoft's monopolistic practices in defense of its operating system, Deleuze and

Guattari associate the arborescent system with "the radicle solution, the structure of Power" (17). In contrast they suggest "acentered systems, finite networks of automata in which communication runs from any neighbor to any other, the stems or channels do not preexist, and all individuals are interchangeable, defined only by their *state* at a given moment—such that the local operations are coordinated and the final, global result synchronized without a central agency" (17). The World Wide Web, viewed with browsers that are multi-platform and free to the public, describes an "acentered system" that thrives on communication from one "neighbor" to any other and eschews the intrusion of governmental regulation.[11]

Nor is there any hierarchy to the nodes on the World Wide Web. As a system of information exchange it was created without institutional ownership or central agency. Only a modifiable agreement on the protocol of exchange exists.[12] The first condition of cyberspace, its decenteredness, appeals to Western individualism and the resistance to political and intellectual forms of totalitarian order. The second condition—a more or less benign quasi-anarchy—extends the promise of an apparently limitless freedom. But the third condition, the always incomplete status of the information grid, may strike some clients of the Web as more bothersome. No keyword search is ever complete; no public newsgroup ever reaches consensus or an end to its deliberation; the contents and location of anyone's web page are never definitive. The decentered, quasi-anarchic, and radically incomplete because inexhaustible universe of cyberspace demands a textual strategy that even a literary genius of Dantesque proportions would be inadequate to describe. Like failed attempts to write the Great American Novel in the earlier part of this century, the experience of cyberspace could never be expressed by a single author. Only the interrelation of multiple authors, a hypertextual and simultaneous transmission of many works in diverse styles, would approach the *élan* of the information age.

As the author of encyclopædic texts impressive for their range of knowledge, Thomas Pynchon would seem to have a special purchase on the problems of writing fiction in the information age. In an essay published in *The New York Times Book Review* in 1984, Pynchon asks the question, "Is It O.K. to Be a Luddite?" The author of *The Crying of Lot 49* (in which Oedipa Maas labors to comprehend how Maxwell's Demon violates the second law of thermodynamics by a reversal of entropy) and of *Gravity's Rainbow* (with its references to cybernetics, ballistics, the calculus, and the chemistry of a new class of plastics known as aromatic polymers) reassesses C(harles). P(ercy). Snow's famous essay, "The Two Cultures and the Scientific Revolution." In 1959 Snow decried the schism between the "literary" and "scientific" intellectual communities such that neither faction would or could understand and appreciate the work of the other. Snow judged literary intellectuals who had ignored the Industrial Revolution and its technological repercussions to be "natural Luddites," after the group of early nineteenth-century British agrarians and laborers who destroyed machines associated with the textile industry. Pynchon, who entered Cornell University as a scholarship student in engineering physics and graduated in 1959 with a degree in English, and who subsequently worked as a technical writer at Boeing Company in Seattle, testily rebukes Snow for casting the problem in rigid binary terms—in effect for thinking too much like a scientist—because such a polarization between "literary" and "scientific" thought no longer applied in the postmodern age. Pynchon argues, "Since 1959, we have come to live among flows of data more vast than anything the world has seen. Demystification is the order of our day, all the cats are jumping out of all the bags and even beginning to mingle. We immediately suspect ego insecurity in people who may still try to hide behind the jargon of a specialty or pretend to some data base forever 'beyond' the reach of a layman. Anybody with the time, literacy, and access fee can get together with just about any piece of specialized knowledge s/he may need. So, to that extent, the two-cultures quarrel can no longer be sustained. As a visit to any local library or magazine rack will easily confirm, there are now so many more than two cultures that the problem has really become how to find the time to read anything outside one's own specialty" ("Luddite" 1). Forty years have passed since Pynchon received his Bachelor of Arts degree and Snow published his essay. (Not to villainize Baron Snow of Leicester [1905–1980], in addition to being a physicist and statesman, he wrote an eleven-volume novel series, *Strangers and Brothers* [1940–1970], that analyzes power and the relation between science and the community while delineating the changes in English life in the twentieth century.) We're now in a postindustrial age in which the major technological advances are forged in silicon rather than steel. Swift and easy access to data in an intellectual climate that recognizes and encourages multidisciplinary projects has healed some of the rift between the "literary" and "scientific" communities, as the fiction that I wish to discuss will illustrate. The problem, as Pynchon rightly appraises it, is not the exclusivity of fields of knowledge but the sheer volume of information available to those who would earnestly pursue it.

The crisis of technology and industrial production in the first half of the twentieth century has been transformed into a millennial crisis of information and communications technology. In his fiction Pynchon has always expressed deep ambivalence toward technology. His accounts of mathematical and scientific discovery in *Gravity's Rainbow* reveal an intellectual fascination with what can be accomplished through technology. But the Luddite in Pynchon worries that technology—especially pertaining to nuclear weaponry—also represents an inescapable threat to the human condition. The computing machines on which we now rely have made themselves indispensable; they represent a concentration of capital that far exceeds the worth of their human attendants. Technology in the control of the military-industrial complex or the multinational corporation can be profoundly anti-humanistic. It is for these aggregate bodies of power, wealth, and authority that Pynchon reserves his greatest ire and suspicion, rather than for technology itself.

Pynchon concludes his essay by stating that we live "in the Computer Age" in which there "seems to be a growing consensus that knowledge really is power, that there is a pretty straightforward conversion between money and information" ("Luddite" 41). The Luddite insurrection was primarily an uprising of an exploited class of laborer against the crushing domination of industrialized capital.[13] Within a year after implementing the assembly-line technique of mass production in 1913, Henry Ford, in an attempt to assuage the fears of labor, announced a profit-sharing plan for his 13,000 employees, who received a minimum wage of $5 per eight-hour day. Responding to an inquiry from a *New York Times* correspondent, Ford denied that he was a socialist.[14] But in postindustrial terms the confrontation appears to be not one of labor versus capital but between specialist and non-specialist. Each professional discipline generates so much and such complex data that it takes a specialist to reconnoiter the field and gather the relevant data. Although any one of us may be specialists in our field—endocrinology, Lacanian psychoanalysis, computer programming—we become the latter-day equivalent of informational Luddites when faced with the expertise and guidance required to approach another field. For Pynchon, hope for the non-specialist has "come to reside in the computer's ability to get the right data to those whom the data will do the most good." It's a nice thought, but one detects a dollop of irony when he concludes, "With the proper deployment of budget and computer time, we will cure cancer, save ourselves from nuclear extinction, grow food for everybody, detoxify the results of industrial greed gone berserk—realize all the wistful pipe dreams of our days" ("Luddite" 41). The question for paranoids is, if anyone can get access to the data, how do you know they'll do something decent with it?

An article by Nancy Jo Sales, "Meet Your Neighbor, Thomas Pynchon," originally published in *New York* magazine, contains a photograph of what one can only presume is Pynchon and his son Jackson walking (back to the camera) down a Manhattan street. The essay bills itself as a "literary investigation." Some might call it a literary "outing" of one of the publishing world's most notorious recluses. Or one can more skeptically view it as a form of advance publicity for Pynchon's much anticipated "big" novel, *Mason & Dixon*, published by Henry Holt in April 1997. Though it verges on literary gossip, this article presents an interesting counterpoint to Pynchon's Luddite position. As many know, Pynchon has been strongly averse to publicity. The last verified photograph to which he voluntarily submitted was taken in 1955;[15] he does not make public appearances; and *People* magazine called Pynchon's address "the best-kept secret in publishing." How then did Sales and the intrepid staffers of *New York* track Pynchon down when so many other literary sleuths had failed for years? Sales reveals her source:

> On an openly accessible online service that uses a cross-referencing of credit-card and telephone numbers, *New York* discovered Thomas Pynchon's address in about ten minutes. "We have recently moved," Pynchon noted in his introduction to *Slow Learner*, a book of early short stories, "into an era when . . . everybody can share an inconceivably enormous amount of information, just by stroking a few keys on a terminal."
>
> With just a few keystrokes, *New York* was able to learn that Pynchon's neighborhood [in Manhattan] is a bustling, civil, and prosperous one, with good subway access. . . . The neighborhood is New York at its finest: tolerant, navigable, sane. It's the New York that could almost lull one into believing that broader American myth, the one Pynchon warns against in his fiction: All is right with the world. (62)

Information is power. An individual, even without corporate sponsorship, can break down dense barriers, some of which have been maintained for a long time. Computer access to vast networks of information can be incredibly liberating for the individual: she can assimilate, cross-reference, and synthesize data; she can gain access to databases that might allow her to challenge the assumptions of dominant cultural institutions; she can create and distribute information so swiftly and so broadly that she defies more traditional loci of dissemination such as publishers, universities, advertising agencies, news organizations, or political institutions. The layperson with relatively modest resources can compete with the specialist whose work is most often endorsed and funded by institutions. But if we take Nancy Jo Sales's investigative claims literally, it's obvious that the information explosion also represents one of the greatest threats to individual privacy. Data regarding individuals passes beyond their control and can be manipulated by other persons or institutions by the same means that prove so liberating.

Pynchon's regard for the computer as an information resource oscillates between an anxiety that the compilation of enormous databases will be used by panoptic institutions to restrict individual freedom and enforce social patterning, and the hope that a "counterforce" of technological savants will challenge the hegemony of the oppressor. His swift betrayal by the journalistic corps, however, suggests that the author-as-recluse, commenting from the secure margins of mass culture, has become an anachronism in the media age with its cult of instant celebrity. The unreachable, sacerdotal, and autonomous author has fallen casualty to new modes of communication. Recall the first Proverb for Paranoids in *Gravity's Rainbow:* "You may never get to touch the Master, but you can tickle his creatures" (237). Pynchon cannot expect to address the crises of an information society and remain himself untouched by its media arm. There has been a shift from the introspective privacy of reading to the public interactions of hypermedia and the collaborative authoring of documents. A lifetime might be recorded in the form of sealed letters that under penalty of federal law only the addressee may open; now the flurry of e-mail messages that are transmitted and stored in the public domain may be intercepted by one's employer or a governmental agency. One observes that the literary author who wishes to retain his privacy—Pynchon, Gaddis, Salinger, DeLillo—has to a large extent been eclipsed in the publishing industry by authors who move effortlessly in the public domain as media commodities.

The Superabundance of Information

An examination of the work of several postmodern novelists shows that they consider science and technology stimulating intellectual and cultural resources for fiction. But the information explosion presents a special challenge to a discipline whose commitment to text is patient, meticulous, and time-consuming. Before turning to specific textual strategies for dealing with the superfund of information, let me outline some problems that digital culture presents for the book-bound. The immediate quandary in the age of multimedia is superfluous choice. The ridiculous assortment of electronic media available for our entertainment and assistance—cellular phones, cable and satellite television (now with so many offerings one needs a program to review the programming), multimedia personal computers (that in Windows 98 features desktop "channels")—practically enforces channel surfing. This phenomenon doesn't represent discontent or boredom with what one's viewing, so much as the anxiety that—with so many choices—there could be something more compelling elsewhere. The attention fragments, and superfluous choices lead to no-choice. In the multi-tasking environment one completes all of nothing. Surplus experience in postindustrial society results in boredom and anxiety. The world deprives us of an intensive experience by offering too much.[16] Rather than compete directly with the hyperextension of choice in electronic media, reading should offer an intensive, absorptive experience. Jameson relates this disjunctive experience of postmodern culture to schizophrenia "in the form of a rubble of distinct and unrelated signifiers" (*Postmodernism* 26). As a palliative to the displacement and superficiality of television, fiction has the capacity to convey complex ideas, psychological and emotional states, and even a density of informational content that the ever-flickering visual stimuli of electronic media cannot.

When researching any field or topic of interest, one encounters the superabundance of data. Advances in the technology required to generate, compress, sort, store, and retrieve information have contributed to the exponential increase in the quantity of information available to anyone who searches for it. The limited capacity of our attention and memory is easily overwhelmed. The profusion of data compels the human mind to relinquish its function as a repository of knowledge to the machine. The mind acts as the principal agent of the search strategy, in the synthesis of diverse data retrieved, and in the analysis of results. But all of these intellectual activities have become increasingly reliant upon a technological interface in order to comprehend—both in the sense of to understand and to gather—information. For the novel to have any cultural impact in the twenty-first century, it must act upon those aspects of the intellect that are most exercised by the information age. The romantic appraisal of the author as creative personality should give way to the functions of selection and organization of the already copious materials available for inclusion in a work of fiction. Even more so than in the metafictive works of Barth, Barthelme, Hawkes, and Coover, the novel at the millennium should aspire to the apprehension of form rather than the retention of content. Rather than replicate the formulae of popular genres, fiction should recombine, invent, and mutate genres, including forms of nonfiction. Instead of following in the same linear and episodic paths, fiction should strategize new, nonlinear narrative structures. And the novel must be the imaginative realization of cultural forms that have meaning and relevance.

The information age confronts us all—the littérateur and the general reader alike—with the superabundance of information. The on-line ecology fosters a remarkably low ratio of signal to noise; or, to put it another way, the Internet promulgates an enormous quantity of data that's of little consequence to anyone. Donald Barthelme ironically anticipates this condition in his novel *Snow White* (1967) in which the seven dwarves consider the escalation of per-capita production of trash. They decide that eventually "the question turns from a question of disposing of this 'trash' to a question of appreciating its qualities" (97). In fact they prefer reading books that "have a lot of *dreck* in them, matter which presents itself as not wholly relevant (or indeed, at all relevant) but which, carefully attended to, can supply a kind of 'sense' of what is going on" (106). Though we take the literary pronouncements of the dwarves as tongue-in-cheek, still it becomes the obligation of the novelist to provide discernment or filtering of the diminished percentage of culturally relevant information and the critical savvy to express the "sense" that can be derived from a body of work that passes into total irrelevance. Such attention to the local scale of relevance is also necessary on the global scale. Information —to the extent that it is not noise—will always display some sort of patterning. But information in large quantities may have no discernible organization. Again, to quote Barthelme's mischievous dwarves—who are in turn making a mockery of Henry James, "Where is the figure in the carpet? Or is it just . . . carpet?" (129; Barthelme's ellipsis).

Simulated Cyberfiction

Let's turn, finally, to an examination of some techniques that postmodern fiction adopts in addressing an information-saturated age. William Gibson's *Neuromancer* (1984) introduces the concept of "cyberspace," a form of virtual reality in which the human consciousness is transported through and interacts with fields of data. The term "cyberspace" has achieved common parlance, but with a definition limited to the transfer of data in a network, as in this *Washington Post* editorial of 1997 that notes, "business people with an interest in cyberspace can be heard compiling year-end figures and asking one another why more people aren't transacting their financial business over the Internet." I'm sure that the protagonist of Gibson's novel, Case, who is a twenty-first century "cowboy," or data thief, would consider filching credit-card numbers off the Internet mere child's play. His ambition is to be "jacked into a custom cyberspace deck that projected his disembodied consciousness into the consensual hallucination that was the matrix" (*Neuromancer* 5). He exults in the conceptualized transfer of his being, not bytes. Gibson actually wrote his novel on a mechanical typewriter. He tells cyberpunk critic Larry McCaffery in 1986, "It wasn't until I could finally afford a computer of my own that I found out there's a drive mechanism inside—this little thing that spins around. I'd been expecting an exotic crystalline thing, a cyberspace deck or something, and what I got was a little piece of a Victorian engine that made noises like a scratchy old record player" ("Interview" 270). Disenchanted by these vestiges of the Industrial Revolution, Gibson neuro-romanticizes the disembodied, digital Information Age. In his cyberspace, human consciousness and the computer matrix, "bright lattices of logic unfolding across that colorless void" (*Neuromancer* 5), coexist or "dance" in the same dimension of reality.

In the same interview, Gibson declares that "Information is the dominant scientific metaphor of our age, so we need to face it, to try to understand what it means. It's not that technology has changed everything by transforming it into codes. Newtonians didn't see things in terms of information exchange, but today we do" ("Interview" 273). As information becomes the only currency, control of information becomes the ultimate power. In the "Sprawl" or BAMA, the megalopolis created by the Boston-Atlanta Metropolitan Axis (*Neuromancer* 43), it is "difficult to transact legitimate business with cash" (6). The written word has become an obsolete and inessential form of communication (88). And virtual wars have been fought over access to the computer nexus, with national and mercantile mainframes defended by ICE, or Intrusion Countermeasures Electronics (28). *Neuromancer* envisages a transformation of information from a passive commodity into an active force, from content that is traded into a substantive medium through which one moves.[17] By the end of the novel (whose plot I won't try to summarize here), a second transformation takes place. The informational matrix becomes an autonomous entity; i.e., the matrix assumes an ontological status independent of, and far more powerful than, the systems that created it. Late capitalism has already made us familiar with the multinational corporations represented in the novel by the CHUBB financial system or the Tessier-Ashpool core. But *Neuromancer* concludes with the merger—or perhaps apotheosis—of two Artificial Intelligences, Wintermute and Neuromancer. Their union, as the narrator takes pains to make clear, is comparable to the joining of the two hemispheres of the cerebral cortex, the left associated with verbal skills and logical functions, and the right hemisphere associated with the perception of nonverbal visual patterns and emotion. "Wintermute was hive mind, decision maker, effecting change in the world and outside. Neuromancer was personality. Neuromancer was immortality" (269). Together they form an unnamed, virtually transcendent informational entity. In its own words, the AI is "Nowhere. Everywhere. I'm the sum total of the works, the whole show" (269). The awakening of information to itself thus extends beyond human cognizance or the ability of human technology to control.

Neuromancer represents a technological dystopia. It has relatively little in common with classical science fiction. Gibson says, "When I write about technology, I write about how it has *already* affected our lives; I don't extrapolate in the way I was taught an SF writer should. . . . My aim isn't to provide specific predictions or judgments so much as to find a suitable fictional context in which to examine the very mixed blessings of technology" ("Interview" 274).[18] He borrows at will from the conventions of genre fiction, and cuts across several cultural levels in his allusions. His hero, Case, is closely modeled on the *film noir* detective; he's an independent contractor ultimately working in opposition to the multinational mainframe—a counter-cultural figure. But there are also progressive aspects of technology presented in the novel. In *Neuromancer* and its sequels, *Count Zero* (1986) and *Mona Lisa Overdrive* (1988), characters can "jack in" to a "simstim" (or simulated stimulation) deck that allows them a surrogate physiological experience.[19] George Landow and Paul Delany, developers of hypermedia programs, note that the "total reproduction of experience" in simstim has already been modeled "in the world of entertainment and for military and commercial training" (8). In literary studies, however, they anticipate that hypermedia, an electronic text with video and audio links, won't reach the sensory-tactile stage until well into the next century. Digital replication is an extension of the technology of compact disks. Reproduction (of an individual, a work of art, or a stimulus) implies a relation of original to copy. But in digital replication, it becomes impossible to distinguish between copies. Everything is a simulacrum, and the "genuine" has no meaning. Formulated by Jean Baudrillard as the "precession of simulacra," a copy without an original (or "master" in recording industry terms), the simulation is "inaugurated by a liquidation of referentials. . . . It is no longer a question of imitation, nor duplication, nor even parody. It is a question of substituting the signs of the real for the real" (*Simulacra and Simulation* 2). The totalitarian implications of replication are answered in *Neuromancer* by "dub, a sensuous mosaic cooked from vast libraries of digitalized pop; it was worship . . . and a sense of community" (104). The digitized environment can be either exactly replicated or radically scrambled.

Gibson attends to the cybernetic adage of Marshall McLuhan that "the medium is the message." His conceptualization of "cyberspace" and "simstim" —the quicksilver environment through which his characters move—necessitates a prose style and narrative technique that are equally accelerated. McCaffery praises Gibson for his "technopoetic prose surface" and "reliance on the cut-up methods and quickfire stream of dissociated images characteristic of William S. Burroughs and J. G. Ballard" (*Storming the Reality Studio* 263–64). George Slusser attributes a new style to cyberpunk literature that emulates the speed of electronic communication and the immediacy of the visual image: "In . . . Gibson's *Neuromancer* (1984), a new style does operate, a mode that is to traditional narrative as MTV is to the feature film. Images have been condensed, sharpened, creating an optical surface—a matrix of images that is more a glitterspace, images no longer capable of connecting to form the figurative space of mythos or story. This is optical prose, one more proof that the printed word, as McLuhan suggests, has succumbed to the fragmenting speed, the instantaneity and monodimensionality of the visual image" (McCaffery, *Storming* 334). In addition to the acceleration granted by the profusion of visual montage, one also recognizes the effects of immediate relocation made possible by the shift from analog to digital technology. One is able to skip instantaneously from one location to another and shuffle the order of play in digitally recorded media, whereas in the analogue medium of the cassette or reel-to-reel tape one must proceed in a linear and continuous fashion until reaching the desired location, hampering the speed with which one may effect any transition. The narrator of *Neuromancer* explains that "Cyberspace, as the deck presented it, had no particular relationship with the deck's physical whereabouts" (105). Case "jacks in" and "jacks out" of his cyberspace deck, and there are abrupt shifts in the narrative scene as he "flips" between his own embodied experience and the simstim link with Molly. In *Mona Lisa Overdrive* such transitions occur with "Not even a click" (229), no servo-mechanical auditory cue at all. In the case of Slick Henry, whose short-term memory was compromised while serving hard time in a chemo-penal unit, "Transition like that scares the shit out of him" (227). The narration in cyberpunk fiction switches without warning between virtual and actual modes of reality, between replicated memory and present action, between holographic and physiological experience, contributing to the dislocation of the reader. Gibson himself acknowledges that "I recognized that cyberspace allowed for a lot more *moves*, because characters can be sucked into apparent realities—which means you can place them in any sort of setting or against any backdrop. In some ways I tried to downplay that aspect, because if I over-did it I'd have an open-ended plot premise. That kind of freedom can be dangerous because you don't have to justify what's happening in terms of the logic of character or plot" ("Interview" 272–73). While Gibson's self-restraint is appreciated by his sometimes puzzled readers, it's the risks he takes in venturing into the nonlinear, decentered, and open-ended narrative space that make the novel both exciting and revolutionary. As Slusser suggests, "Literature cannot use traditional techniques to present a contemporary reality because that reality has been transformed by technical advance to a point where those techniques no longer fit it. . . . For the cyberpunkers, it is the information age and the increasing fusion of electronic matrix and human brain" that demand new literary techniques (McCaffery, *Storming* 334). *Neuromancer* theorizes a world in which the instantaneous electronic transfer of consciousness makes the need for physical contact obsolete.

Readers and critics of Gibson's novels have spied evidence of the narrative conventions of popular genres such as the western, the hard-boiled detective fiction, and the military adventure. Like other common narratives, these genres naturalize continuity of scene, consistency of character, and sequential plot. But far from relying on standard literary forms, Gibson pirates them. Just as eighteenth-century buccaneers preyed on ships of all nations, and software pirates today replicate the programs of many commercial vendors, so Gibson plunders indiscriminately from the holds of other literary forms to enrich the cyberpunk novel. His use of multiple literary codes—the appropriation of the conventions of several popular genres within one novel—always foregrounds traits that are familiar to the reader. Such thefts are brazenly self-aware, taking the reader as willing and necessary accomplice in their borrowing. Gibson's piracy places him in league with the eclectic force of postmodernism identified by Charles Jencks.[20] There are allusions to high culture in *Neuromancer*—Marcel Duchamp's *Large Glass* rests in the possession of the Tessier-Ashpool estate (207); and 3Jane expresses her contempt for her autocratic father by echoing Sylvia Plath's "Daddy" (219). But more pronounced is the pluralist, hybridized, cross-cultural coding that lends the book its prickly resistance to convention—the "vatgrown ninja assassin" (74), a Rastafarian ganja-smoking orbital pilot (220), and the branding of microprocessor implants with the Mitsubishi-Genentech corporate logo (10). For stealing from a single genre one might indict Gibson for banality; but in pirating multiple source-codes and fashioning them into the cyberpunk program, Gibson must be acquitted of the charge that he unselfconsciously employs cartoon characters and timeworn plots. In fact, one should add Japanese *anime* to the list of popular sources for his work. Gibson's eclecticism only provides further evidence of postmodern innovation in the novel.[21]

Along with Bruce Sterling, Gibson is considered the originator of the "cyberpunk" movement in contemporary science fiction. But his emphasis on computer-mediated experience and information technology has had an important influence on a range of postmodern writers. One notable protégé—although that couldn't be the correct term for so rebellious an author—is Kathy Acker, who incorporates elements of punk, gothic horror, raw pornography, and feminist theory in her fiction. Acker's novel *Empire of the Senseless* (1988) pays homage—again, not quite the right word—to *Neuromancer* by plagiarizing two extended passages from it. In an essay that relates cyberpunk and postmodern writing, "POSTcyberMODERNpunkISM," the critic Brian

McHale shows that Acker "appropriates and rewrites material" from Gibson's novel (*Constructing Postmodernism* 233).[22] The female ninja Molly, who is Case's ally and lover in *Neuromancer*, appears to be an important model for the female protagonist Abhor in *Empire of the Senseless*. Molly has a surgically inset visor and retractable four-centimeter scalpel blades under her burgundy fingernails (*Neuromancer* 24–25). Abhor is described as "part robot, and part black" (*Empire* 3) and a cybernetic personality "construct" (34). McHale points out that the worlds of both *Neuromancer* and *Empire* are in post-apocalyptic ruin "dominated by omnipotent multinationals and intelligence organizations" (*Constructing Postmodernism* 235). But Acker's reworking of Gibson's materials has neither the ironic intention of parody nor the dishonest intention of plagiarism (a theft of intellectual property with the hope that the reader will not recognize the source text). In fact, Acker practices a form of "sampling" or "dub" (made possible by the digital replication and splicing of audio-files) that assumes the intended audience will not only recognize the source but appreciate how its insertion in a new context alters its qualities and meaning. In shifting attributes from Molly to Abhor, Acker critiques the fantasies of cyborg sex and violence that are rather tightly laced in *Neuromancer* and mocks the adolescent male subjectivity that comprises the largest audience for science fiction. Abhor's cyborg-minority status also underlines the contempt that a technologically dominant patriarchy retains over women, half-breeds of any sort, and other politically oppressed groups. The technique of sampling represents an innovation that the ease of transferring text in an electronic medium has brought to postmodern fiction. Michael Joyce, hypertext novelist and co-author of the writing program *StorySpace*, remarks that various kinds of electronic text, including e-mail, hypertext nodes, and word-processed documents, can be moved nomadically and iteratively from one location, occasion, or perspective to another. He contends that the "nomadic movement of ideas is made effortless by the electronic medium that makes it easy to cross borders (or erase them) with the swipe of a mouse, carrying as much of the world as you will on the etched arrow of light that makes up a cursor" (*Of Two Minds* 3). As Acker's appropriation of Gibson indicates, postmodern fiction has already instituted the practice of sampling and shifting text as the compact disk replaces the bound book and electronic text replaces the printed page.

McHale suggests that Acker may have been intrigued by Gibson's establishing of a "simstim" link between Case and Molly, so that he could vicariously experience and monitor her progress as she breaks into a corporate headquarters to steal an electronic storage unit that contains the taped memory and personality-construct of a dead hacker. "Case's access to the sensory experience of his female partner" creates the "potential for gender disorientation" (*Constructing Postmodernism* 234). As anyone who has participated pseudonymously in chat rooms, or Multiple User Dungeons, on Internet services knows, gender, race, age, and appearance can all be altered, constructed, or exchanged in virtual reality. There are several exchanges of gender orientation and sexuality throughout *Empire of the Senseless,* beginning with the opening section in which Abhor speaks in the first person about her childhood through the identity of her male lover, Thivai, whose parentage in turn is either alien or robotic (155). From Acker's anti-hegemonic and feminist perspective, the ambiguity and alteration at will of identity is politically liberating. Acker's fiction exemplifies several of the theoretical positions expressed by Donna Haraway in her essay, "A Cyborg Manifesto: Science, Technology, and Socialist-Feminism in the Late Twentieth Century." Haraway defines the "cyborg" as a "cybernetic organism, a hybrid of machine and organism, a creature of social reality as well as a creature of fiction" (*Simians, Cyborgs, and Women* 149). As such the cyborg is a "creature in a post-gender world" (150). Acker's "constructs" (another borrowing from Gibson in *Empire of the Senseless* [34]) represent a feminist self that is neither biologically nor culturally determined. Her cyborgs transgress boundaries "between mind and body, animal and human, organism and machine, public and private, nature and culture, men and women, primitive and civilized" (*Simians, Cyborgs, and Women* 163). In the information system described by Haraway, the biological function of reproduction cedes to an artificial practice of replication. Gibson projects the dystopian consequences of cloning in *Mona Lisa Overdrive,* in which the Tessier-Ashpool alliance is insured by the cryogenic storage of ten identical sets of male and female children, despite laws "governing the artificial replication of an individual's genetic material" (125). But whereas the study of literary "influence" inevitably invokes the patriarchal myth of reproduction —of a father and the dilution of his gene pool among his descendants—the relationship of Acker to Gibson might be more effectively viewed as the artifice of replication, and hence a relationship of equivalence. "In the beginning was the copy" (Haraway 216). Acker and Gibson as novelists are postmodern replicants.

Two recent novels by Richard Powers are occasionally misclassified as science fiction when in fact they are fictions about science. Powers risks disappointing fans of the genre by wholly avoiding any element of "future shock" on the presumption that addressing cutting-edge scientific theory is suitably complex and stimulating to the imagination. His ideal reader is that rarest of hybrids who devours his monthly copy of *Scientific American,* has a remarkable memory for passages in *The Norton Anthology of English Literature,* Volumes I and II, and for reasons relating to Powers's expatriate sojourn to the Netherlands, has a strong interest in the painting of the Dutch masters. Powers's books are clearly not intended for anyone with a contempt for erudition. Each chapter tests the reader with literary quotations, cultural allusions, and scientific references. *The Gold Bug Variations* (1991), at 638 pages in length, is an entrant in the class of encyclopædic narrative. Like Pynchon's *Gravity's Rainbow,* it attempts to render the full range of knowledge that would be deemed essential for an educated individual in the present state of the culture, including aspects of the physical sciences.[23] About this novel, Powers says "I wanted to write the encyclopedia of the Information Age" (*Galatea* 215). The female protagonist of the novel is naturally a reference librarian, Jan O'Deigh, for whom sorting, gathering, and relating information is a profession. The title of the novel weaves references to J. S. Bach's *Goldberg Variations* (an Aria, followed by thirty variations and an Aria da capo, for the harpsichord [1742]), E. A. Poe's detective tale of encryption "The Gold-Bug" (1840), and the genetic variations encoded in the Watson-Crick model of the double-helix of DNA (1953). Just as Bach's *tour de force* is based on "levels of pattern, fractal self-resemblances" (*Gold Bug* 579) among its base line, canonical variations, and structure, so Powers's novel adopts the intricate 32-part form in its Aria (appropriately, a poem), thirty chapters, and a recapitulation. The literary equivalent of Douglas Hofstadter's *Gödel, Escher, Bach: An Eternal Golden Braid,* Powers's novel makes the argument that the complex formulation of the human mind, its artful creations, and the structures of the physical world are not only comparable in worthiness of study but more closely related in form than is commonly accepted. As the first-person narrator (who happens to be named Richard Powers) of *Galatea 2.2* (1995) exclaims, "Where the hell is the Two Cultures split when you need it?" (44). Powers has accepted the challenge of C. P. Snow to bring the realms of the sciences and the humanities into accord, recognizing the powerful imaginative requirement of scientific discovery and the deeply rational nature of the arts.

The plot of both novels depends on an acquaintance with information theory. In *Gold Bug,* a specialist in information science pursues an enigmatic molecular biologist, Stuart Ressler, who set out in 1957 to crack the genetic code. Ressler and other geneticists puzzle over the four nucleotides that offer a "seductive suggestion of encoded information assembling the entire organism" (*Gold Bug* 44). In *Galatea,* the "author" joins forces with a cranky cognitive neurologist named Philip Lentz whose project is to model the human brain by means of a computer-based neural network. In the field of connectionism, "Neural networkers no longer wrote out procedures or specified machine behaviors. They dispensed with comprehensive flowcharts and instructions. Rather, they used a mass of separate processors to simulate connected brain cells. They taught communities of these independent, decision-making units how to modify their own connections. Then they stepped back and watched their synthetic neurons sort and associate external stimuli" (*Galatea* 14). Lentz and Powers endeavor to train successive generations of their neural net, eventually dubbed Helen, to analyze a reading list of Great Books. As cybernetic Pygmalions, they employ "repeated inputs and parental feedback" (15) to reinforce associations. The governing of a young mind also requires a selective sheltering from the worst of human vice and misery by her instructors, and Helen is accordingly kept unaware of and unconnected to the world beyond her circuitry until the end. In contrast to the intimate conversation that develops between trainer and artificial intelligence, Powers describes the World Wide Web as "yet another total disorientation that became status quo without anyone realizing it." He finds that the web "seemed to be self-assembling" and facilitates the linking of "endless local investigations" like ice crystals on a pane of glass (7). Yet the web overwhelms him.[24] It is finally a failure, rather than a success, of communication: "The web was a neighborhood more efficiently lonely than the one it replaced. Its solitude was bigger and faster. When relentless intelligence finally completed its program, when the terminal drop box brought the last barefoot, abused child on line and everyone could at last say anything instantly to everyone else in existence, it seemed to me we'd still have nothing to say to each other" (9). Despite being a massive parallel processor at the supercomputing facility of a thinly disguised University of Illinois at Urbana-Champaign, Helen thrives on the cultural coherence of literary works; she is subject to, and subject of, great books. The web has achieved total connectivity, in the opinion of the author-narrator, at the cost of a locatable and sympathetic identity. While Helen emulates human consciousness, the web—"a vast, silent stock exchange trading in ever more anonymous and hostile" missives (9)—threatens humane characteristics. The Turing test comes when Helen competes with a twenty-two-year-old master's candidate in English in a comprehensive examination. The young student, however, challenges the literary canon as a measure of cultural knowledge when she asks, "Has [Helen] read the language poets? Acker? Anything remotely working-class? Can she rap?" (284). The questions, of course, are self-directed ones for Powers. Although his ambitiously intellectual novels bridge the gulf between literature and science, he endorses neither Acker's attack on the canon and patriarchal discourse in novels such as *Great Expectations* nor Gibson's enthusiasm for the disembodied consciousness of cyberspace.

Since his first novel, *Americana* (1971), Don DeLillo's subject has been the superfluity, the self-absorption, and the paranoia of American culture. Readers of DeLillo are familiar with his often satiric assessment of the preprocessed, disposable, filtered, and homogenized material of the American environment. A special target has been the dulling uniformity of media culture, the talking heads that pronounce on the day's events in unaccented newsspeak in thousands of identical motel rooms across the country. His brilliant novel *White Noise* (1986) serves as a catalogue of the information age's malaise. As a visiting lecturer in popular culture at the College-on-the-Hill, Murray Siskind tries to persuade his colleague Jack Gladney that the "medium" of television "is a primal force in the American home. Sealed-off, timeless, self-contained, self-referring. It's like a myth being born right there in our living room" (51). The hypnotic blue fire emanating from the television set becomes iconic, not the mere channel through which information flows but an object of reverence and compulsion. The immanent presence of the media—suffused throughout and saturating the environment—encases the viewer's sensorium in data like an artificial myelin sheath.

Underworld (1997) is DeLillo's most monumental work of fiction, demanding comparison with recent blockbusters of literary fiction such as Pynchon's *Mason & Dixon* (also published in 1997), David Foster Wallace's *Infinite Jest* (1996), William H. Gass's *The Tunnel* (1995), and Joseph McElroy's *Women and Men* (1986). Tom LeClair, one of DeLillo's most devoted critics, has called these works "prodigious fictions," in part for their daunting length (ranging from 650 to over a thousand pages in the case of the novels by Wallace and McElroy), and in part because they are conceived as "information systems, as long-running programs of data with a collaborative genesis" ("Prodigious" 14). One might expect that in a media economy of "sound bite" politics, frantic pitchmen, and the jump-cut editing made fashionable by MTV videos, such maximal-sized publications would meet the fate of other leviathans. In fact the publication of an anthology called *Sudden Fiction* (short stories of remarkable brevity, with a preface by Robert Coover), would seem to fit the mood of the age much better. When the attention span of audiences dwindles to no more than a few seconds, what is one to make of this cluster of excessively long novels? Is this the work of novelists writing in spite of distracted or nonexistent readers? DeLillo has written a "novel that tries to be equal to the complexities and excesses of the culture," which he described in an interview given while *Underworld* was in process ("Art of Fiction" 290). The superabundance of information, in Baudrillard's phrase "diffused, and diffracted in the real" (*Simulacra* 30), would overwhelm a merely anecdotal representation or cross-section of the culture of excess. *Underworld* takes as its initial condition the headlines from the front page of *The New York Times* for

October 4, 1951 that appear in symmetrical juxtaposition: to the left, "the Giants capture the pennant, beating the Dodgers on a dramatic home run in the ninth inning. And to the right, symmetrically mated, same typeface, same-size type, same number of lines, the USSR explodes an atomic bomb" (668). America's entrance into the Cold War, with its pervasive fear of nuclear annihilation, haunts the entire novel. The confetti that falls on the field of the Polo Grounds (described in the first section of the novel, "The Triumph of Death") is doubled by the human ashes that drifted down onto Hiroshima. In a character doubling that is appallingly apt for the nuclear age, the novel presents J. Edgar Hoover (who shares a celebrity box at the playoff game with Frank Sinatra, Toots Shor, and Jackie Gleason) and the Catholic nun Sister Edgar as celibate hypochondriacs. They take it as their special mission to instill the morbid fear of imminent nuclear vaporization. Sister Edgar checks to see that each student wears an identification tag during schoolroom duck-and-cover drills. Reviewer James Wood complains that the novel's "big, broken structure moves back and forward through all the decades since 1951, travelling from Arizona to the Bronx to Kazakhstan and back again, and the book intends to be a collection of lavish fragments, set down in a maze" ("Black Noise" 38). *Underworld*, however, aspires to an historically capacious form by moving in nonlinear fashion from one rather precise, critical moment in the postwar American experience through the unreliable and unpredictable texture of history without attempting to impose a singular "story" on events (or the fictional and historical characters implicated in those events). DeLillo's paranoid investigations into American history in *Underworld* take the form of a "strange attractor" that plots a path unpredictably folded over and through itself without ever exactly repeating its path. Such a nonlinear divagation through history requires a maximal length in which to express the system to its fullest.

As a teenager in the Bronx, Nick Shay "takes his radio up to the roof of his building so he can listen alone, a Dodger fan slouched in the gloaming" (32) to the legendary playoff game on October 3, 1951 between the Brooklyn Dodgers and the New York Giants at the now demolished Polo Grounds.[25] Baseball fans recall that Bobby Thomson's ninth-inning home run off Dodger pitcher Ralph Branca, called "The Shot Heard Round the World," capped an improbable stretch run by the Giants to overtake the Dodgers and win the pennant. *Underworld* follows the path of that baseball—which becomes another strange attractor within the novel—as it leaves Branca's hand, sails over the forlorn head of Dodger outfielder Andy Pafko at the wall, into the stands and into history. First held by a turnstile-jumper from Harlem, Cotter Martin,[26] sold to advertising executive Charles Wainwright for $32 and change, passed to his ne'er-do-well son Chuckie, a navigator on a B-52 making bombing runs over Vietnam, tracked down by a collector of baseball memorabilia, Marvin Lundy, the scuffed baseball is finally acquired by Nick Shay as an expiation for various losses in his life. Lundy explains that the problem with the Thomson home-run ball is its provenance, to use the art world's term for an object's record of possession. He admits, "'I traced it all the way back to October fourth, the day after the game, nineteen hundred and fifty-one. . . . I don't have the last link that I can connect backwards from the Wainwright ball to the ball making contact with Bobby Thomson's bat.' He looked sourly at the scoreboard clock. 'I have a certain number of missing hours I still have to find'" (180–81). Lundy's inability to establish the exact "lineage" of the scuffed baseball holds more significance than setting its commercial value in the "nostalgia trade." The baseball represents the problem of provenance in American culture at large. It's impossible to trace American culture to some authoritative source. All along the way one encounters fakes, pretenders, and simulations. As Baudrillard remarks, "Disneyland exists in order to hide that it is the 'real' country, all of 'real' America that *is* Disneyland" (*Simulacra* 12). DeLillo suggests that America is a thing made of plywood, like the replica of the Polo Grounds' scoreboard in Lundy's basement, painted over to resemble a civilization. The counterexample in the novel is Sabato Rodia's Watts Towers in Los Angeles, a sculpture "rucked in the vernacular" that still achieves an "epic quality," constructed by an Italian immigrant over many years using "whatever objects he could forage and scrounge" (492). DeLillo's *Underworld* creates a mosaic from the found materials of American history from the 1950s through the early 1990s with such invention that these details finally achieve an epic proportion.

DeLillo's protagonist, Nick Shay, is an expert in waste management. He confronts the basic crisis of our disposable consumer economy: what to do with the inexhaustible mountains of toxic garbage generated every day. Nick believes that "waste is a religious thing. We entomb contaminated waste with a sense of reverence and dread. It is necessary to respect what we discard." Although his job forces him to consider the collective refuse of American culture, the great impasto that binds us all together as a nation, Nick's upbringing still shapes his approach to the problem. "The Jesuits taught me to examine things for second meanings and deeper connections. Were they thinking about waste? We were waste managers, waste giants, we processed universal waste" (*Underworld* 88). Again speaking of the Disneyland that is America, Baudrillard discovers "a space of the regeneration of the imaginary as waste-treatment plants are elsewhere, and even here. Everywhere today one must recycle waste, and the dreams, the phantasms, the historical, fairylike, legendary imaginary of children and adults is a waste product, the first great toxic excrement of a hyperreal civilization" (*Simulacra* 13). Whereas Jack Gladney in *White Noise* sought through the superabundance of noise for hidden messages, scraps of meaning, Nick Shay sifts through the mountainous remains of American culture for a few slips of sublime dream-life that might give his existence meaning.

The stunning epilogue to the novel, "Das Kapital," catapults the reader from the nostalgia of newspaper headlines, the nascence of television, and the ethnic neighborhoods of the Bronx in the 1950s into a postindustrial society dominated by "the flow of information, through transnational media, the attenuating influence of money that's electronic and sex that's cyberspaced, untouched money and computer-safe sex, the convergence of consumer desire" (*Underworld* 785). As peripatetic as Nick Shay is, flying to Kazakhstan, Zurich, and Lisbon in the 1990s to confront "the intractability of waste" (805) as a disposal expert in the service of multinational capitalism, he still experiences a profound uneasiness with the virtuality of postmodern hyperspace that Jameson suggests has "finally succeeded in transcending the capacities of the individual human body to locate itself, to organize its immediate surroundings perceptually, and cognitively to map its position in a mappable external world" (*Postmodernism* 44). Although Nick finds some "sense of order and command reinforced by the office" and its "contact points" of telecommunications (*Underworld* 806), he longs for "the days of disorder . . . when I was alive on the earth, rippling in the quick of my skin, heedless and real. I was dumb-muscled and angry and real" (810). As a product of the 1950s, Nick touts visceral contact over disembodied connections, real streets over hyperlinked systems, and grappling hotly with disorder over the cool regard of cybernetic order. His slacker son, Jeff, however, is a pure product of the information age, "a lurker. He visits sites but does not post. He gathers the waves and rays. He adds components and functions and sits before a spreading mass of compatible hardware. The real miracle is the web, the net, where everybody is everywhere at once, and he is there among them, unseen" (808). Jeff discovers the web page devoted to miracles and containing an account of the South Bronx wild-child Esmeralda. A web site that chronicles miracles—ironic as the preservation of blueprints for the atomic bomb by post-apocalyptic monks in Walter Miller's *A Canticle for Leibowitz* (1972)—supplants the power of faith in the supernatural with "the power of false faith, the faith of paranoia" in the unseen connections of the web. The web offers no antidote to the grim but real underworld of the South Bronx. But in cyberspace "Everything is connected. All human knowledge gathered and linked, hyperlinked, this site leading to that, this fact referenced to that, a keystroke, a mouse-click, a password—world without end, amen" (825). As in so many other epics there is the lure of a false paradise that offers repose to the weary, food to the hungry, and pleasure to those who have experienced privation. All of which precede loss of purpose, loss of direction, and loss of identity. For DeLillo cyberspace is another mirage, a lotusland: "Is cyberspace a thing within the world or is it the other way around? Which contains the other, and how can you tell for sure?" (826).

The superabundance of information overwhelms the reader because of the impossibility of assimilating the retrievable data, or because of the difficulty of distinguishing essential from inessential data. Postmodern fiction offers certain palliatives for these symptoms of technological neurasthenia. Novels are the tangible manifestation of intellectual and imaginative competence. They are complex systems of thought created by an individual and realized in the mind of another individual without reference to a user's manual. They are the cybernetic threshers of meaning. So in the universe of information, where true significance appears as sparsely distributed as interstellar dust, the novels of William Gibson, Kathy Acker, Thomas Pynchon, Don DeLillo and Richard Powers register as gravitational events whose concentration of matter curves space around them. Finally, postmodern fiction offers relief for the "pixelated,"[27] those viewers stunned into anomie by the bombardment of pixels—the smallest image-forming units of the video display. It turns out that print on paper still has the capacity to evoke images and ideas as compelling as any we might encounter in the flicker of a screen.

Notes

Chapter 1. Being in Uncertainties

1. Lyotard's proposition of a postmodern science has drawn the ire of some scientists and science writers who are reluctant to consign determinism, and its applicable presence in both advanced research and our everyday lives, to the hopper. Critiques are levied by Alan Sokal and Jean Bricmont in *Fashionable Nonsense,* in which they argue that Lyotard provides insufficient evidence for his claims of the limits of predictability (134–38). Although he doesn't mention Lyotard by name, John Horgan, in his forebodingly titled *The End of Science,* attacks an "ironic science" that too much "resembles literary criticism in that it offers points of view, opinions, which are, at best, interesting, which provoke further comment. But it does not converge on the truth. It cannot achieve empirically verifiable surprises that force scientists to make substantial revisions in their basic description of reality" (7).

2. The most useful historiographic renditions of postmodernism are Anderson, *The Origins of Postmodernity,* Bertens, *The Idea of the Postmodern: A History,* and Hutcheon, *The Politics of Postmodernism.* For a more dialectical reading of postmodernism and its predecessors, see Hassan, *The Postmodern Turn.*

3. Robert McAlmon's memoir of American expatriates in Paris during the 1920s, *Being Geniuses Together,* gives some indication of the wide appeal of the genius in modernism. Bob Perelman's study of difficulty and illegibility in modern literature, *The Trouble with Genius,* focuses on how the cult of genius drives modernist writers such as Ezra Pound, Joyce, Stein, and Louis Zukofsky to produce ambitious writing structures and strategies.

4. Woolf is of course heralding the arrival of modernism as the successor to realism in "Mr. Bennett and Mrs. Brown." Although she remarks the change was not as sudden as a hen laying an egg, nevertheless she observes, "All human relations have shifted. . . . And when human relations change there is at the same time a change in religion, conduct, politics, and literature. Let us agree to place one of these changes about the year 1910" (92). Jencks marks the birth of postmodernism, and the symbolic "death of Modern architecture" (*Post-Modernism* 27) and its ideology of progress on the date of demolition of the Pruitt-Igoe housing project. The demise of the Pruitt-

Igoe, which was built in the International style of modernist architecture, represents the failure of Le Corbusier's social program to bring order to low-income neighborhoods through architectural planning and austere structures. To my taste, July 1972 seems a little late for postmodernism's break with its predecessor. Positing an earlier break, one could read the demolition of Pruitt-Igoe as a rear-guard action, rather than as a revolutionary statement.

5. Hassan examines the impact of the New Science on modernism, arguing that "relativity, uncertainty, complementarity, and incompleteness are not simply mathematical idealizations; they are concepts that begin to constitute our cultural languages; they are part of a new order of knowledge founded on both indeterminacy and immanence" (55–64).

6. In a letter to Ronald Lane Latimer (15 November 1935) Stevens says, "In 'The Idea of Order at Key West' life has ceased to be a matter of chance. It may be that every man introduces his own order into the life about him and that the idea of order in general is simply what Bishop Berkeley might have called a fortuitous concourse of personal orders. But still there is order. This is the sort of development you are looking for. But then, I never thought that it was a fixed philosophic proposition that life was a mass of irrelevancies any more than I now think that it is a fixed philosophic proposition that every man introduces his own order as part of a general order. These are tentative ideas for the purposes of poetry" (*Letters* 293). Stevens discounts the theorics of "bishop's books" in part three of "Connoisseur of Chaos" (*Collected Poems* 215). In another formulation of this essentially modernist belief, Roland Barthes argues that "the work of art is what man wrests from chance" (*Critical Essays* 217).

7. In his own footnote to lines 367–77 of the poem, Eliot quotes a passage from Hermann Hesse's *A Glimpse into Chaos:* "Already half Europe . . . is on the road to chaos. . . . The bourgeois laughs, offended, at these songs, the saint and the prophet hear them with tears" (*Complete Poems* 54). One can infer that Eliot—as both saint and prophet—shares Hesse's lamentation at the rise of political disorder and the chaos of war.

8. The Viking Critical Library edition of *Portrait* provides the source of Joyce's allusion to Flaubert in his letter to Mlle. Leroyer de Chantepie (537).

9. On the tendency of science to emphasize soluble equations and stable systems in textbooks and relegate disorderly, unstable systems to the realm of perversion, James Gleick comments, "most differential equations cannot be solved at all. . . . Only a few [scientists] were able to remember that the solvable, orderly, linear systems were the aberrations. Only a few, that is, understood how nonlinear nature is in its soul" (67–68).

10. The intractability of nonlinear equations in physics courses, and in the textbooks that guide their progress, often results in their relegation to the very end of discussion, as "special cases." As Hayles says, "The practice reinforces the assumption, implicit in Newtonian mechanics and encoded within the linguistic structure of stem and prefix, that linearity is the rule of nature, nonlinearity the exception. Chaos theory has revealed that in fact the opposite is true" (*Chaos Bound* 11).

11. In "Fictions as Dissipative Structures: Prigogine's Theory and Postmodernism's

Roadshow," David Porush aligns the "dissipative systems" that Prigogine claims are capable of bringing order out of chaos with the cluster of automotive, road novels in postmodernism. The flow of traffic and the flow of information obey similar rules of behavior.

12. "Chaosmos" is James Joyce's portmanteau word for the interrelation of chaos and cosmos, disorder and order. For a discussion of the symbolic order that underlies the surface chaos of *Ulysses*, see Umberto Eco, *The Aesthetics of Chaosmos*. Thomas Jackson Rice argues that Joyce's texts demonstrate "the emergence of complex narrative behaviors, arising from simple initial premises" (1), thus anticipating the fields of chaos and complexity in his modernist fiction. Although I am inclined to sympathize with a reading in which the practices of fiction precede comparable discoveries in the natural sciences, I find that Joyce's residual commitment to symbolic order in *Ulysses* (as a kind of initial premise) is in keeping with the modernist compulsion to an orderly centering rather than a postmodern inclination to disorderly dispersal. Ultimately, analyses of Joyce's "chaosmos" and "chaoplexity" will focus on the interrelation of their component behaviors.

13. Hayles finds "isomorphisms" between poststructuralist literary theory and complex dynamics, but she does not assume that such similarities are "the result of direct influence between one site and another. In particular, I am *not* arguing that the science of chaos is the originary site from which chaotics emanates into the culture. Rather, both the literary and scientific manifestations of chaotics are involved in feedback loops with the culture" (*Chaos and Order* 7).

14. For excellent discussions of Cage's music and writing within the context of chaotics, see Hayles, "Chance Operations: Cagean Paradox and Contemporary Science," and Retallack, "Poethics of a Complex Realism."

15. In "Back from Chaos," the sociobiologist E. O. Wilson advances a thesis of "consilience" or the unity of science that would return the natural sciences, social sciences, and humanities to a shared core of beliefs that has been lacking since the Enlightenment, whose thinkers "knew a lot about everything." As I am suggesting here, Wilson states that specialists, thinkers who know a lot about a little, gain in prominence during the period of modernism: "Fragmentation of expertise was furthered in the twentieth century by modernism in the arts, including architecture. The work of the masters—Braque, Picasso, Stravinsky, Eliot, Joyce, Martha Graham, Gropius, Frank Lloyd Wright, and their peers—was so novel and discursive as to thwart generic classification" (56). In contrast to my proposal that the pluralist freedom of postmodernism invites interdisciplinarity, Wilson, more conservatively to be sure, argues that the radical skepticism of postmodernism (associated with deconstruction) is "the ultimate antithesis of the Enlightenment" (58), and defeats any synthesis of knowledge. Wilson's leap "back from chaos" and other theories of indeterminacy suggests a blindness to the very arena in which the arts and the sciences find common discourse, though not the unity (on the ground of the natural sciences) that he would prefer.

16. Snow proposes that the cultural divide—although apparent in all Western industrialized nations—was exacerbated in England by "our fanatical belief in edu-

cational specialisation" and the more rigid "social forms" of the English educational system (18).

17. Murray Gell-Mann adopts "quark" after browsing through Joyce's *Finnegans Wake*, perhaps in appreciation of Joyce's whimsy. But Gell-Mann assigns a referent to the word that it couldn't carry in its original context, where it appears as a toast, in the first line of a song, "—Three quarks for Muster Mark!" (383).

18. To his credit Snow recognized the preponderance of men among his colleagues in physics, and argues that it is "one of our major follies that, whatever we say, we don't in reality regard women as suitable for scientific careers" (57). On the continuing difficulties faced by women in scientific careers, even in postmodernism, see Haraway, *Simians, Cyborgs, and Women*.

19. Coincidences are always meaningful in Pynchon. The first sputnik satellite was launched by the Soviet Union on October 4, 1957. Snow's Rede Lecture, meant to encourage curriculum reforms and spur further training of scientists, was published two years later. Pynchon received his bachelor's degree in the spring of 1959; his essay is published in October, 1984. He is a *sputnik zemlyi*, a fellow traveler.

20. Hayles's study, *The Cosmic Web: Scientific Field Models and Literary Strategies in the Twentieth Century*, addresses the importance of field theory to in literary texts by Lawrence, Nabokov, Borges, Pirsig, and Pynchon. She makes no mention of the role of composition by field in the open-form poetics of the Black Mountain School.

21. For a debate between Horgan and Kauffman on the legitimacy of complexity theory and the imminence of the end of science, see the electronic journal *Hotwired* (the on-line version of *Wired* magazine), in its feature "Brain Tennis" (with exchanges between participants logged via e-mail over several days). It can be accessed at <http://www.hotwired.com/braintennis/96/25/indexoa.html>.

22. Hayles points out that there is little mention of Prigogine's work in irreversible thermodynamics and "self-organization" in Gleick's *Chaos*, and conversely Prigogine makes no mention of Mandelbrot's fractal geometry and other figures critical to Gleick's history of chaos theory (*Chaos and Order* 12–13). In *Chaos Bound* she suggests the difficulty lies in the differing mathematical techniques that each branch uses to analyze chaos (10). But there appears to be a methodological disparity as well, in that the strange-attractor branch has concentrated on experimental results whereas the emergent-order branch has made more of the philosophical implications of their research. See also Porush, "Making Chaos: Two Views of a New Science."

23. Rice discusses the presence of both an imbedded or immanent design, and an emergent order, in the composition of Joyce's *Ulysses* (106–109). At the same time Rice wishes to identify Joyce as a "scientific realist" whose methods (more) accurately represent nature in its complexity. He attempts to dissociate chaos theory (and Joyce) from the indeterminacy of postmodernism and poststructuralism, in opposition to Hayles (*Chaos Bound*) and Argyros.

24. See Paulson on the relation of order and complexity in terms of information theory. He opposes both order and complexity to chaos defined as random disorder. "Order" is "characterized by a high degree of redundancy and thus a low level of

information." "Complexity" is "characterized by a low level of redundancy and high degree of information. The complex state may be formally identical to the chaotic state but with the following crucial difference: its particular arrangement is meaningful or important in some context" (72).

25. For much more extensive analyses of the constraints imposed by Perec on the composition of *Life*, see Harris, Magné, and Perloff, "The Return of the (Numerical) Repressed."

26. LeClair discusses Gaddis's *J R* as an example of a systems novel that runs to excess and aspires to mastery (*Art of Excess* 87–105).

Chapter 2. Design and Debris

1. For a thorough analysis of *Travesty* in relation to Poe, see Charles Berryman, "Hawkes and Poe: *Travesty*." Charles Baxter's fine essay, "In the Suicide Seat: Reading John Hawkes's *Travesty*," also identifies trauma as "the central experience of the novel" (874). It is also amusing to note that the driver finds the ticking of the dashboard clock to be "unbearable" (*Travesty* 34), reminiscent of Poe's "The Tell-Tale Heart."

2. In another anecdote, Hawkes also reveals some identification with Henri's terror: "I soon developed a horror of automobiles through trying to drive an ambulance" in the American Field Service in World War II; with little driving experience, "I ran the ambulance into a boulder and blew out a tire" ("A Conversation" 165–66).

3. Hawkes's insistence on the "literal accident" distinguishes his work from later cyberpunk fiction, such as Neal Stephenson's *Snow Crash* (1992), with its proposition of a virtual reality that is equal to but other than the physical world. One can compare Hawkes's driver in *Travesty* to the protagonist of J. G. Ballard's *Crash* (1973), also named Ballard, since they share a streak of sadomasochism. Just as Ballard becomes erotically obsessed with deliberate, suicidal automobile accidents, the driver in *Travesty* may seek more than jealous revenge on Henri, Chantal, and Honorine; he may take sexual pleasure in the impact itself. My experience in teaching *Travesty* has been that students react with visceral repulsion at the very thought of the crash—a reaction that Hawkes would not wish to deprive his readers of by consigning the novel to mere fantasy. Similarly, David Cronenberg's controversial film adaptation of Ballard's *Crash* (1996) sickened some viewers and was banned in Britain. For an extended discussion of vehicular mayhem and chaos theory in postmodern fiction, especially William Marshall's *Roadshow* (1985), see Porush, "Fictions as Dissipative Structures."

4. The driver here attempts to refute the accusation of depravity that might be attributed to him, much as Herman Melville attributes it to Claggart in *Billy Budd, Sailor*: "Though the man's even temper and discreet bearing would seem to intimate a mind peculiarly subject to the law of reason, not the less in heart he would seem to riot in complete exemption from that law, having apparently little to do with reason further than to employ it as an ambidexter implement for effecting the irrational. That is to say: Toward the accomplishment of an aim which in wantonness of atrocity

would seem to partake of the insane, he will direct a cool judgment sagacious and sound" (*Novels and Tales* 1383). The driver argues that his reasoned inquisition is no mere mask for an irrational jealousy.

5. For a discussion of the role of chance in the postmodern aesthetic, especially in the novels of William Burroughs and the music of John Cage, see Christopher Butler, *After the Wake: The Contemporary Avant-Garde* (102–108).

6. In *The Noise of Culture: Literary Texts in a World of Information*, William R. Paulson distinguishes between chaotic disorder and complexity: "The terms *order* and *complexity* may lead to confusion, especially in the context of information theory. We can imagine two completely different types of organized states, both of which would be intuitively opposed to chaos or random disorder. The first, which may be appropriately called 'ordered,' is characterized by a high degree of redundancy and thus a low level of information. . . . The second, which would be better called 'complex' or 'varied,' is characterized by a low level of redundancy and a high degree of information. The complex state may be formally identical to the chaotic state but with the following crucial difference: its particular arrangement is meaningful or important in some context" (72). For the driver, the "tableau of chaos" obviously represents a "complex" arrangement.

7. In *Connoisseurs of Chaos: Ideas of Order in Modern American Poetry*, Denis Donoghue comments regarding this poem: "The singer imposes upon reality her own imagination, until reality is taken up into her song and there is nothing but the song. And when the song is ended, the observers . . . find that even in their own eyes reality is mastered, more orderly; the sea and the night are fixed and disposed. . . . By common agreement the sea is reality, things as they are that come to us without invitation or apology, often to be thought of as chaos" (190). Thus the function of the imagination for Stevens is to master chaos, not to illustrate or extol it.

8. In *The Postmodern Turn*, Ihab Hassan describes, albeit in a list of binary oppositions, the distinctions between modernism and postmodernism such that the modernist impulse toward Creation/Totalization is contrasted with the postmodernist emphasis on Decreation/Deconstruction. He ascribes to postmodernism "a vast will to unmaking" (91, 92).

9. As a further example of a dualistic opposition in which artistic ordering fails to maintain itself against encroaching chaos, Henderson proposes that Hawkes's novels express "a dialectic between the art of fictional ordering and the chaos of a 'reality' which is potentially subject to the ordering process but which ultimately threatens to dissolve such designs as are present in created or fictional reality" (3034A). The novelist as "creator" inevitably wages a losing battle against the oppressive forces of chaos.

10. Fractals are frequently illustrated by the Koch "snowflake." See Gleick (98–100), and Hayles (*Chaos Bound* 166–67).

11. For a discussion of "Chinese Box Worlds" in postmodern fiction, see Brian McHale's descriptive catalog in *Postmodernist Fiction* (112–30).

12. Papa denies that his "brain is sewn with the sutures of [Henri's] psychosis" (120), but then reveals that he carries a scrap of paper on which he has written two lines of Henri's poetry, claiming that "I might even have written them myself" (127).

Patrick O'Donnell provides an extensive discussion of the likenesses established between Papa and Henri, as well as those between Chantal, Papa's mistress Monique, and Honorine (*John Hawkes* 139–40).

13. Earlier commentators on Hawkes's novel have noted the source. Paul Emmett points out that "a reviewer here in Chicago contended that the travesty of the title is a highly deliberate and sophisticated travesty of the work and death of Albert Camus. There are, it seems, a number of parallels between *Travesty* and the car ride that resulted in Camus' death" ("A Conversation" 166).

Chapter 3. Discipline and Anarchy

1. In an interview with Larry McCaffery, Acker attributes her introduction to the theories of Gilles Deleuze, Félix Guattari, and Michel Foucault, by Sylvère Lotringer, as providing her with "a greater degree of control and precision about what I was doing. By the time of *Empire of the Senseless* I could even *plan* things! Whereas before I never even wanted to touch anything that was rational, because I felt that would intrude on whatever was going on" ("Path of Abjection" 26). As a result, *Empire* is one of the first novels in Acker's *oeuvre* of non-narrative fiction to show evidence of plotting or consistent characterization.

2. Discussing *Empire* with McCaffery, Acker agrees that she sought a more "constructive" approach to the problems of the patriarchy, to produce "a myth that people could live by." Here she reiterates her three-part plan for the novel, in both deliberate and affirmative terms: "What was involved in the writing itself was a dialectic between trying to get to another society and realizing you can't. So in the first part of the book I basically took the world of patriarchy and then killed the father on every level I could imagine. Part of this involved my attempt to find a way to talk about taboo—those basic transgressions patriarchy is responsible for but tries to cover over and deny. In the second part of the book I tried to imagine a society where you didn't have any of these taboos because it wasn't defined by Oedipal considerations. But even though I wanted to get to that world, I couldn't get there. . . . And so in the last section, there's only the sense of, 'Well, what the fuck do we do now?' That section is the synthesis of what you have in the first two parts of the dialectical movement: the wanting to get to that society of taboo and freedom from the central phallus, but realizing this is impossible. Finally *Empire* isn't just about getting to another society (which is a literal impossibility) but about searching for some kind of myth to replace the phallic myth" ("Path of Abjection" 31–32).

3. Brian McHale carefully examines Acker's appropriation and rewriting of material from Gibson's *Neuromancer* as a form of blank parody distinct to postmodernism in "POSTcyberMODERNpunkISM" (*Constructing Postmodernism* 225–42). McHale's essay is included with modification in McCaffery, ed., *Storming the Reality Studio* (308–23).

4. In his discussion of the "antihegemonic" tendencies of Acker's fiction, "Expectations of Difference: Kathy Acker's Regime of the Senseless," Joseph Natoli pursues a similar line of argument regarding her refusal of clarity, cogency, and orderly behav-

ior. Acker creates "an empire of the senseless only because her constructions are wary of sense and its empire, only because sense is reward for following an order already established to communicate sense. A formation of her own signs, a putting an ending to her own following of that arrangement of signs she is already producing *within* as well as already produced *by*, and a devising in a way that the reader begins to construct and not to follow, a creation of desire so as to uncode desire—this is Acker's empire of the senseless" (*Mots d'Ordre* 140–41).

5. Arthur F. Redding offers a more thorough evaluation of the interplay of sadomasochism and disgust in "Bruises, Roses: Masochism and the Writing of Kathy Acker."

6. For a further disquisition on the Algerian revolution by Acker, see her *Algeria: A Series of Invocations Because Nothing Else Works* (1984). The historical source for Acker's Mackandal can be found in C. L. R. James, *The Black Jacobins*. James identifies Mackandal as a maroon, or runaway slave, from Guinea who organizes a slave revolt in Haiti between 1751 and 1758. A one-armed chieftain, Mackandal leads raids on the French colonists. James observes that an "uninstructed mass, feeling its way to revolution, usually begins by terrorism, and Mackandal aimed at delivering his people by means of poison. . . . He arranged that on a particular day the water of every house in the capital of the province was to be poisoned, and the general attack made on the whites while they were in the convulsions and anguish of death" (20–21). Mackandal's (eventually failed) slave revolt serves as model for the fall of Paris to the Algerians in Acker's *Empire*. In Haitian Voodoo (also alluded to in Acker's text), Mackandal is a loa, or supernatural being invoked by impassioned worshippers, who is manifest as a giant mosquito (Mackandal was captured and burned at the stake, and his followers believed he had metamorphosed into that plaguing insect). In keeping with Acker's characterization, he represents the spirit of revolution on Haiti, he abhors any form of slavery, and he punishes slave traders in a variety of terrible ways. "Agouné" is a loa manifest as the Sea-Master. In *Empire of the Senseless*, Agone (Abhor's partner) is a Cuban-born sailor (119). Acker's further adventures in Haitian voodoo appear in the first work collected in *Literal Madness, Kathy Goes to Haiti* (1988).

7. Greg Lewis Peters asserts in "Dominance and Subversion in Kathy Acker" that "Acker takes [Hélène] Cixous's concept of 'writing the body' very seriously indeed. Her texts are attempts to make the abstract physical through ('through' meaning literally in one side and out the other) the body, giving a visceral form to the feminine writing hypothesized by Cixous and Luce Irigaray" (150). Likewise, Ellen G. Friedman in "'Now Eat Your Mind': An Introduction to the Works of Kathy Acker" argues that for Acker "the body, particularly the female body, becomes the site of revolution. In this regard, Acker, perhaps more directly than many other women writers, creates the feminine texts hypothesized by Hélène Cixous" (39). The tattoo thus becomes the central figure equating the body with the text.

8. This rather Foucauldian passage may be an instance of Acker's infamous "plagiarism." As she remarks in her interview with Friedman, "I did use a number of other texts to write [*Empire*], though the plagiarism is much more covered, hidden. Almost

all the book is taken from other texts" ("Conversation" 16). Michel Foucault, of course, has had much to say about language and the prison. One passage from "The Discourse on Language" seems directly relevant to Acker's concern with taboo: "In a society such as our own we all know the rules of *exclusion*. The most obvious and familiar of these concerns what is *prohibited*. We know perfectly well that we are not free to say just anything, that we cannot simply speak of anything, when we like or where we like; not just anyone, finally, may speak of just anything. We have three types of prohibition, covering objects, ritual with its surrounding circumstances, the privileged or exclusive right to speak of a particular subject; these prohibitions interrelate, reinforce and complement each other, forming a complex web, continually subject to modification. I will note simply that the areas where this web is most tightly woven today, where the danger spots are most numerous, are those dealing with politics and sexuality. It is as though discussion, far from being a transparent, neutral element, allowing us to disarm sexuality and to pacify politics, were one of those privileged areas in which they exercised some of their more awesome powers. In appearance, speech may well be of little account, but the prohibitions surrounding it soon reveal its links with desire and power" (*Archaeology of Knowledge* 216).

9. In "Writing, Identity, and Copyright in the Net Age," Acker suggests that in the "free zone" of the Internet, "*copyright* as we now define it will become a thing of the past" (*Bodies of Work* 101). Recent copyright infringement cases pitting the music industry against Napster and MP3.com suggest that this conflict is far from resolution. But Acker is certainly astute when she suggests that the "work isn't the property, it's the copyright" ("Devoured" 12); it's not the material, it's the code of law that defines the work in capitalism. For further discussion of plagiarism in Acker, see Naomi Jacobs, "Kathy Acker and the Plagiarized Self," and Arthur Saltzman, "Kathy Acker's Guerilla Mnemonics."

10. Peters observes in "Dominance and Subversion," "In *Empire,* Abhor . . . is imprisoned linguistically and sexually as well as literally. She is eventually freed from literal prison to form her own, one-cyborg motorcycle gang. In this world without patriarchy and taboo, she creates chaos by attempting to drive on the highway according to the rules of the Highway Code, computing stopping distances and measuring speed while other vehicles crash all around her. She is metaphorically learning the codes of a language that has no semantics to accommodate her. Just as Acker explodes patriarchal language by reinventing/plagiarizing its sacred texts, so Abhor's actions reveal the fundamental uselessness of any male code to express a specifically female experience. That both Acker and Abhor reject the spirit of the codes while working within the letter of them is one more form of capitulation, but a reasoned capitulation that is subversive in intent" (154–55).

11. One wonders just what it means to ride a human by the rules. What equipment would be called for? What injuries might result?

12. For the emergence of order in systems far from equilibrium, see Prigogine and Stengers, *Order Out of Chaos*.

13. As an indication that Acker is enamored with the figure of Rimbaud, she

includes a fictionalized biography of the symbolist poet, "Rimbaud," as the first section of *In Memoriam to Identity* (1990).

14. For an interesting application of the principle of the nomad and smooth space in Deleuze and Guattari to Acker's earlier novel, *Don Quixote* (1986), see Douglas Shields Dix, "Kathy Acker's *Don Quixote:* Nomad Writing."

Chapter 4. American Oulipo

1. The remark appears as the conclusion to an editorial in *The Blind Man* (May 1917), "The Richard Mutt Case," in defense of Duchamp's infamous readymade, *Fountain*, itself a piece of plumbing culled from the J. L. Mott Iron Works in March, 1917. Steven Watson points out that the editorial was jointly composed by the New York Dadaists, including Beatrice Wood. See *Strange Bedfellows* (281, 318).

2. During the Depression, the Works Progress Administration (WPA) and the Federal Arts Project commissioned a survey of American decorative arts and crafts, the *Index of American Design*. Many artists and writers, including the poet Louis Zukofsky, contributed to this monumental project. See Nadel, "A Precision of Appeal."

3. For a far better introduction to and history of the Oulipo, with more detailed discussions of their works than I can provide here, see Motte.

4. The design for "The Castle" appears on p. 40, and that for "The Tavern" on p. 98. For an extended discussion of Calvino's proceduralism in *The Castle of Crossed Destinies*, see my essay, "The Uncertain Predictor: Calvino's *Castle* of Tarot Cards."

5. Louis Mackey states that Sorrentino uses the Rider deck, designed by Edward Waite and drawn by Pamela Coleman Smith, though he provides no discussion of why Sorrentino chose this particular tarot deck (87). Perhaps it is more available to Americans, and more modern, than the Marseilles or Visconti decks.

6. As I mentioned earlier, there is a deliberate imperfection in Perec's design, because he skips the bottom left-hand square (1,0) between chapters 65 and 66. Bellos discusses this and other aspects of Oulipian constraint in his article, "Georges Perec's Puzzling Style." The figure of the apartment building appears on p. 501 of *Life*.

7. According to William McPheron, Sorrentino's bibliographer, the working title of the novel was *Triad*. McPheron identifies the allusion to Flann O'Brien in his background notes for *Odd Number* (*Gilbert Sorrentino* 63). Though the student author in *At Swim-Two-Birds* echoes him, it is the Pooka who declares that "truth is an odd number" (155). See also Cowley, "Gilbert Sorrentino," for discussion of these sources in Sorrentino's novels.

8. See Motte's discussion of tradition and experiment in Le Lionnais, Jean Lescure, and Jacques Bens in his Introduction to *Oulipo* 6–9.

9. Motte refers to the "seeming paradox . . . in Oulipian aesthetics, the belief that systems of formal constraint—far from restricting a writer—actually afford a field of creative liberty" (*Oulipo* 18).

10. Gleick describes the arithmetic process in greater detail, but my point rests on the relation of simple rules and complex products. He says, "A Mandelbrot set program

needs just a few essential pieces. The main engine is a loop of instructions that takes its starting complex number and applies the arithmetical rule to it. For the Mandelbrot set, the rule is this: $z \rightarrow z^2 + c$, where z begins at zero and c is the complex number corresponding to the point being tested" (231). Multiplying complex (imaginary) numbers, however, is not all that simple, and the thousands of iterations required to form a well-defined fractal necessitates a powerful computer.

11. The second informant expatiates that *Mouth of Steel* is a book "in three parts, each one a different kind of part, I mean different stories, sort of, different characters. . . . It's an odd item" (85, 89).

12. Mackey identifies the source of the title, *Isolate Flecks,* in the trilogy and other works by Sorrentino. The phrase appears in William Carlos Williams's *Spring and All,* XVIII: "It is only in isolate flecks that / something / is given off."

13. Brian McHale describes infinite regress as one form of textual infinity found in Jorge Luis Borges, Barth, and Federman in *Postmodernist Fiction* (114–15). Louis Mackey points out that Barth's *Lost in the Funhouse* (1969) begins with a "Frame Tale" that, as per the author's instructions, may be cut out, twisted once, and fastened end to end as a Möbius strip: "Once Upon A Time There Was A Story That Began" (1–2).

14. Henslowe's catalogue of props (Sorrentino retains their Elizabethan spellings) includes Chayne of dragons, Little alter, Payer of stayers, Cage, Two mose bankes, Tree of gowlden apples, Three tombes, Beacon, Baye tree, Rocke, Bedsteade, Chyme of belles, Hell mought, Wooden canepie, and Sittie of Rome.

15. Sorrentino is quite familiar with Jakobson's essays on linguistics and poetics. In *Rose Theatre* Sheila's photographic image is "a figure located somewhere on the axis of selection" (240). For Jakobson's analysis of selection and contiguity and their function in poetry and prose, see *Language in Literature.*

16. "Mathews's Algorithm," originally written in French, has been translated into English and included in Motte, *Oulipo* 126–39.

17. John Barth was born on May 27, 1930, and his character, Simon William Behler, is born on July 1, 1930 (based on information supplied in the First Voyage (*Last Voyage* 28). Barth's middle name, Simmons, is echoed in Daisy Moore's playful nickname for Simon, "Persimmon," or 'Simmon for short. There are, of course, many other parallel biographical details in the novel.

18. In this book of twinnings, both Sindbad's favorite ship, and the cutter sailed from Sri Lanka by Behler and Julia Moore are named the *Zahir.* The Arabic word, Behler explains, refers to an apparently unremarkable object that, once it's caught one's attention, nevertheless becomes an "unexpectedly obsessive image or memory" (*Last Voyage* 328). He also makes literary reference to the story by Jorge Luis Borges, "The Zahir" (*Labyrinths* 156–64).

Chapter 5. Noise and Signal

1. William Gibson attributes the coinage of this cyberspace term to Ted Mooney in his novel *Easy Travel to Other Planets* (1981) in his interview with Larry McCaf-

fery. See "An Interview with William Gibson" in McCaffery, ed., *Storming the Reality Studio* (276–77).

2. In "The Figure in the Static: *White Noise,*" Arthur M. Saltzman suggests that the danger of "sensory overload" to language lies in "technological fallout in all its multifarious forms, including such linguistic manifestations as secret codes, arcana, and all the kabbala of conspiracy" (808).

3. Paul Civello, in his chapter on "Fields, Systems, and DeLillo's Postmodern Transformation of Literary Naturalism" in *American Literary Naturalism and its Twentieth Century Transformations*, describes the "undermining of linear causality" in DeLillo's work. He relates this development to the systems theory of Ludwig von Bertalanffy. Bertalanffy's attack on deterministic causes leads to a concept of history not as "a linear narrative of causes and effects brought about by the actions of 'great men,' but a complex, whirling system of often arbitrary and accidental events that are simultaneously causes and effects of still other events—often actuated by seemingly insignificant figures" (123).

4. For an extended discussion of the relationship between the aesthetics of kitsch and postmodernism see Matei Calinescu, *Five Faces of Modernity* (225–312).

5. In DeLillo's novel, *Underworld*, Brian Glassic, one of several characters employed by a waste management corporation, gazes on the Fresh Kills landfill on Staten Island and imagines "he was watching the construction of the Great Pyramid at Giza. . . . He found the sight inspiring. All this ingenuity and labor, this delicate effort to fit maximum waste into diminishing space" (184). Every culture finds solace and inspiration in its defining icon.

6. In her brief discussion of *White Noise* in "Postmodern Parataxis: Embodied Texts, Weightless Information," N. Katherine Hayles casts Murray Siskind as a sort of Jamesonian analyst of late-capitalist commodification: "Suskind" [*sic*] is not concerned with "material reality but the hidden codes embodied within it. . . . Like the UPC markings on the packages, the meanings encoded into the supermarket itself may be decipherable, given the proper equipment. A separate issue is whether anyone wants to make the effort" (409). Murray regards the modern supermarket as a "revelation" (*White Noise* 38) not offered by the steamy New York delicatessens of his youth. But such an *entrée* into the cultural logic of late capitalism comes at the expense of living in a technologically mediated environment. The meanings of the price codes are no longer apprehensible without technology, the "proper equipment" required to decode them; our senses—alert to the aroma of the deli's corned beef—fail us in an environment constructed by the multinational corporate food industry.

7. Douglas Keesey reports in *Don DeLillo* that DeLillo's working title for the novel was "Panasonic," until the Matsushita Corporation refused permission (xii).

8. Tom LeClair comments, "Although Siskind does offer Jack, as well as the reader, penetrating interpretations of the world, especially the meaning of its communications systems, Siskind's advice promotes a profoundly immoral act [the shooting of Willie Mink]. . . . Siskind is also, in Michel Serres's systemic terms, the 'parasite,' the guest who exchanges talk for food and (simultaneously, in French) the agent of

noise in a cybernetic system" (*In the Loop* 221). "Les Parasites" also denotes "static interference" or "white noise."

9. Murray Siskind's rooming house is located across the street from a psychiatric facility. Gladney inquires, "Do you get any noise from the insane asylum?" (49).

10. The riotously funny passage in which Molloy struggles to find a solution to the permutational process of stones and pockets extends for several pages and is thus too lengthy to quote here. A brief selection, however, suffices to show Beckett's intent to satirize the rational, orderly, and mathematical mind. At one point Molloy determines that his principle of "trim" or the "distribution of the sixteen stones in four groups of four, one group in each pocket, . . . had vitiated my calculations until then and rendered the problem literally insoluble. And it was on the basis of this interpretation whether right or wrong, that I finally reached a solution, inelegant assuredly, but sound, sound" (*Three Novels* 71).

11. For the topological analysis of correlation, and the homogeneity or heterogeneity of space, see Deleuze and Guattari's essay, "1440: The Smooth and the Striated," in *A Thousand Plateaus* (474–500).

12. Jack's lecture on the plot to kill Hitler clearly anticipates the subject matter of DeLillo's next novel, *Libra* (1988), whose protagonist is the assassin Lee Harvey Oswald. Jack intones, "All plots tend to move deathward. This is the nature of plots. Political plots, terrorist plots, lovers' plots, narrative plots, plots that are part of children's games. We edge nearer death every time we plot. It is like a contract that all must sign, the plotters as well as those who are targets of the plot" (26).

13. Pneumoconiosis, a chronic lung disease caused by the inhalation of dust particles from silica, asbestos, iron, coal or other minerals or metals. Black lung disease had been identified as early as the Civil War, and cases were quite prevalent throughout the coal fields of Appalachia in the twentieth century.

14. In his lecture "Society and the Scientific Imagination," neurobiologist Dr. Henri Korn reveals the presence of "synaptic noise" in the brain, the permanent activity of the neurons, which behave in a chaotic and unpredictable fashion. Such uncorrelated behavior of the neurons, intensified by the relentless bombardment of electronic noise in the media environment, might be the cause of "brain fade."

15. Norbert Wiener, in *Cybernetics* (1948) and *The Human Use of Human Beings* (1950), establishes the comparison of communication and control processes in biological and artificial systems, pointing out that "the nervous system and the automatic machine are fundamentally alike in that they are devices which make decisions on the basis of decisions they have made in the past," or Wiener's thesis of "feedback." Thus, "the synapse in the living organism corresponds to the switching device in the machine" (*Human Use* 33–34).

16. Richard Powers's provocative novel, *Galatea 2.2* (1995), combines literary and cybernetic pursuits as a novelist-in-residence at the Center for the Study of Advanced Sciences and a cognitive neurologist combine forces to model the human brain by means of computer-based neural networks. The test of the computer's capacity is whether it can analyze literary texts.

17. In his argument for DeLillo as Romantic visionary, Paul Maltby suggests that the "tenor of this passage is not parodic; the reader is prompted by the analytical cast and searching tone of Gladney's narration to listen in earnest. . . . The passage is typical of DeLillo's tendency to seek out transcendent moments in our postmodern lives that hint at possibilities for cultural regeneration" ("Romantic Metaphysics" 261). But one shouldn't confuse a character's sincere search for meaning with the author's. The fact that Gladney experiences a transcendental moment in the expression of a meaningless commercial icon strongly suggests that DeLillo intends the passage to be taken with the requisite dose of postmodern irony.

18. DeLillo would probably date the onset of American postmodernism from the assassination of John F. Kennedy on November 22, 1963, a moment he treats at length in his subsequent novel, *Libra* (1988). In an interview with Anthony DeCurtis, he suggests "that what's been missing over these past twenty-five years is a sense of a manageable reality. Much of that feeling can be traced to that one moment in Dallas. We seem much more aware of elements like randomness and ambiguity and chaos since then" ("Outsider" 48). See also DeLillo's comments in his "Art of Fiction" interview (299).

19. For further discussion of the contemporary mass media in DeLillo's novels, see John Johnston, "Fictions of the Culture Medium: The Novels of Don DeLillo," in his *Information Multiplicity* (165–205).

20. Maltby analyzes this passage as an example of visionary experience in the Romantic tradition, comparing its terminology to that of Edmund Burke's definition of the sublime. I would agree that Gladney's aesthetic response to the chaotic event pertains to sublimity, but I would not strip the passage of its postmodern irony. Maltby, however, argues that the toxic cloud "may seem like an ironic postmodern version of the sublime object insofar as DeLillo substitutes a man-made source of power for a natural one. Yet Gladney's words emphasize that that power is experienced as a *natural* phenomenon" ("Romantic Metaphysics" 271). Since I understand irony to be created by the disparity between appearance and reality, Gladney's experience of the natural sublime in industrial toxicity would still qualify as ironic. But there can be no argument that the event he confronts, whether natural or man-made, behaves chaotically. Perhaps, then, the postmodern experience of sublimity should be defined as an encounter with chaos in any of its forms.

21. In his study of DeLillo's *oeuvre* through the publication of *White Noise* as an emulation of systems theory, *In the Loop: Don DeLillo and the Systems Novel*, LeClair takes the loop as the defining form of the fiction: "What the Gladneys refuse to accept and what forms the basis for DeLillo's understanding of systemic fact and value is the loop: the simultaneity of living and dying, the inherent reciprocity of circular causality that makes certainty impossible. Their refusal is rooted in mechanistic science, that extension of common-sense empiricism which defines the world as a collection of entities, a heap of things like the Gladney's compacted trash, rather than as a system of energy and information" (226).

22. Jack imagines Mr. Gray as "Gray-bodied, staticky, unfinished. The picture wobbled and rolled, the edges of his body flared with random distortion" (241).

23. In *Don DeLillo,* Douglas Keesey points out that Jack's classes "serve to advance Nazism insofar as they encourage students to respond adoringly to Jack's lectures as if he were Hitler giving a fatally hypnotic speech at a mass rally. . . . Jack's kind of teaching does not lead to the understanding and reasoned rejection of fascism, but to its mindless perpetuation" (134).

24. As a point of contrast to Jack Gladney's sense of benediction by information systems, the conspiratorial figure and founder of Radial Matrix, Earl Mudger, in DeLillo's *Running Dog* remarks, "When technology reaches a certain level, people begin to feel like criminals." The paranoia of constant surveillance breeds a sense of guilt and self-incrimination. "Someone is after you, the computers maybe, the machine-police. You can't escape investigation. The facts about you and your whole existence have been collected or are being collected. Banks, insurance companies, credit organizations, tax examiners, passport offices, reporting services, police agencies, intelligence gatherers. It's a little like what I was saying before. Devices make us pliant. If *they* issue a print-out saying we're guilty, then we're guilty. But it goes even deeper, doesn't it? It's the presence alone, the very fact, the superabundance of technology, that makes us feel we're committing crimes. Just the fact that these things exist at this widespread level. The processing machines, the scanners, the sorters. That's enough to make us feel like criminals. What enormous weight. What complex programs. And there's no one to explain it to us" (93). One wouldn't want to ally DeLillo too closely to either Gladney or Mudger, but it seems that Mudger's sense of technophobia and disquietude, and the express complaint that information technology has become inexplicable to its subjects, is closer to DeLillo's authorial position than the rapture of Gladney in *White Noise.*

Chapter 6. The Perfect Game

1. In an interview with Frank Gado, Coover remarks that the widespread popularity of table-top baseball games had taken him somewhat by surprise: "I had thought it was a private idea, a private book, but I soon learned that sort of activity was rather general. . . . Now they even have computers doing it. They compile tapes with all the parameters—down to factors like temperament and stamina—and they distribute the printouts. It's grown to be quite a network. There is even going to be a convention of baseball parlor game players in Philadelphia. Somehow, these enthusiasts have fostered the hope that I should be the secretary for some such organization" ("Robert Coover" 150). Coincidentally, Paul Auster describes how, as an impoverished writer trying to eke out a living in New York in the late 1970s, he "began devising various get-rich-quick schemes. I invented a game (a card baseball game—which was actually quite good) and spent close to six months trying to sell it" ("Interview with McCaffery" 290).

2. Other critics who discuss the ludic in Coover's *Universal Baseball Association* include Caldwell, who explicitly compares the playing of baseball to the making of fiction (162–63), Schwartz, who analyzes the affinity of postmodern writers for baseball, including Coover (145–48), Berman, who contrasts play and competitive sport, and Wallace.

3. See Coover's "The End of Books" and "Hyperfiction: Novels for the Computer."

4. In his often misunderstood essay, "The Literature of Exhaustion," John Barth states that by "exhaustion" he means "the used-upness of certain forms or the felt exhaustion of certain possibilities—by no means necessarily a cause for despair" (64). Barth would agree with Coover that the age of realism, and of the sort of novels it made possible to write, is over; he goes on to suggest, however, that the death of the novel as a literary form would, paradoxically, be a subject of further consideration for the novel. Thus Barth engages himself self-consciously, metafictively with "novels which imitate the form of the Novel, by an author who imitates the role of Author" (72).

5. In the dedicatory prologue to Cervantes in "Seven Exemplary Fictions," Coover also addresses the present epistemological shift: "We seem to have moved from an open-ended, anthropocentric, humanistic, naturalistic, even—to the extent that man may be thought of as making his own universe—optimistic starting point, to one that is closed, cosmic, eternal, supernatural (in its soberest sense), and pessimistic. The return to Being has returned us to Design, to microcosmic images of the macrocosm, to the creation of Beauty within the confines of cosmic or human necessity, to the use of the fabulous to probe beyond the phenomenological, beyond appearances, beyond randomly perceived events, beyond mere history" (*Pricksongs and Descants* 78).

6. According to the *Columbia Encyclopedia*, "Mah Jongg is a four-handed game, played in many variations throughout China, where it probably originated. The standardized Western version is played with 152 tiles, 108 of which are 'suit tiles' (there being three suits), the remainder representing other symbols, usually of greater value. As in Rummy, the object is to build sets, but the rules governing the distribution and accumulation of tiles are extremely complex."

7. In the final chapter the narrator observes that someone's "discovered that the whole damn structure [of the Universal Baseball Association] from the inning organization up and double entry bookkeeping are virtually identical: just multiply it by twenty-one, the guy claims, and you've got it all. Grim idea" (219).

8. See Andersen (60–61); Cope (38, 50–52); Gordon (40–41); and Hansen, "Dice of God" (54–55). Coover acknowledges, "It suddenly occurred to me to use Genesis I.1 to II.3—seven chapters corresponding to the seven days of creation—and this in turn naturally implied an eighth, the apocalyptic day. Having decided on this basic plan, I read a lot of exegetical works on that part of the Bible in order to find out as much as I could which would reinforce and lend meaning to the division into parts" ("Robert Coover" 149).

9. In *Order Out of Chaos*, Ilya Prigogine and Isabelle Stengers distinguish between those chemical reactions that tend toward a final state of chemical equilibrium, or "attractor" state, and those that involve auto- or cross-catalysis, which develop into "reaction loops" that must be described by nonlinear differential equations (131–37).

10. In his important article, "The Dice of God: Einstein, Heisenberg, and Robert Coover," Arlen Hansen examines the debate between Einstein and Heisenberg precipitated by the Solvay Conference in 1927 regarding the ramifications of the uncertainty principle and the probabilistic knowledge implied by quantum theory, about which Einstein famously remarked, "God does not play dice with the universe."

11. Although not specifically concerned with dynamical systems theory, Arthur Saltzman effectively argues that the *Universal Baseball Association* contrasts two different kinds of balance, static and kinetic. Static balance is "inherent in Damon's perfect game: whereas other records are open-ended achievements . . . a perfect game can only be duplicated." Kinetic balance, represented by the league itself, is a process that "accommodates free will and determinism, flex and form." As opposed to the zero-sum axes recorded in the Win-Loss columns for each season's play, the league's administration represents "the perpetual contest of codification and spontaneity that infuse the game with dynamic tension" ("Magisters Ludi" 64–66).

12. The most famous proposition of a Demon whose actions defy the second law of thermodynamics by sorting the orderly from the disorderly world was made by James Clerk Maxwell in *Theory of Heat* (1871). For a compelling discussion of the Demon's function in relation to chaos theory, see N. Katherine Hayles's chapter, "Self-reflexive Metaphors in Maxwell's Demon and Shannon's Choice: Finding the Passages," in *Chaos Bound* (31–60).

13. For a discussion of dissipative structures, period doubling to chaos, and bifurcations in Coover's *Gerald's Party* (1986), see Bernd Klähn, "From Entropy to Chaos-Theories."

14. About the final chapter of the novel Coover comments, "The design, the structure of the book is so self-revealing—and it's not a gloss on the text from which it borrows its design, in the sense of being a theologian's gloss; it's an outsider's gloss, an ironist's gloss. The idea of an 'eighth' chapter [after the seven days of creation], the potential of it, the wonderful ambivalences implied—all this came to me even before I knew what was going to be in it" ("Interview with Robert Coover" 73).

15. Saltzman, following Claude Lévi-Strauss in *The Savage Mind*, suggests the "distinction between games and ritual: the former is disjunctive, ending in the establishment of difference and inequality, while the latter is conjunctive, devising union, or communion, and a merging of sides in the collective, faithful center" ("Magisters Ludi" 60).

Chapter 7. The Excluded Middle

1. Steven Weisenburger, in *A Gravity's Rainbow Companion*, establishes that Part 1 begins on December 18, 1944, in the Advent season, and that Part 4 concludes

on September 14, 1945, on the Feast of the Exaltation of the Holy Cross (10). There is not, however, universal agreement on the exactitude of these dates.

2. Weisenburger asserts that the "chronology unfolds according to a carefully drawn circular design. *Gravity's Rainbow* is not arch-shaped, as is commonly supposed. It is plotted like a mandala, its quadrants carefully marked by Christian feast days that happened to coincide, in 1944–45, with key historical dates and ancient pagan festivals. The implications of this design are several, and wonderfully complex" (9–10). He concludes, "only gravity's rainbow is arch-shaped; the shape of *Gravity's Rainbow* is circular" (11).

3. I use the term "open work" in the sense developed by Umberto Eco, especially with respect to systems of communication and information theory, in *The Open Work* (44–83). Since *Opera aperta* was first published in 1962, there can be no reference to Pynchon's *Gravity's Rainbow* as an example of the open work, but Eco has much to say on the topic of James Joyce's *magnum opus:* "In *Finnegans Wake* we are faced with an even more startling process of 'openness': the book is molded into a curve that bends back on itself, like the Einsteinian universe. The opening word of the first page is the same as the closing word of the last page of the novel [this is not quite correct, as I have indicated; the last/first sentence of the *Wake* can be spliced together, but it does not overlap as Eco states]. Thus, the work is *finite* in one sense, but in another sense it is *unlimited.* Each occurrence, each word stands in a series of possible relations with all the others in the text" (*Open Work* 10). In my conception, Pynchon's novel fits this description as well, as a work of unlimited semiosis in a finite form.

4. Edward Mendelson suggests that "the only conventionally modernist sections of the book are the Slothrop sequences, with their private point of view and stream of consciousness. Slothrop's disintegration, Pynchon implies, summarizes the historical fate of literary modernism" (166).

5. Or almost no one. The prodigious Dr. Stuart Ressler, molecular biologist and scientific hero of Richard Powers's encyclopedic novel, *The Gold Bug Variations* (1991), is presented with a complete *Encyclopedia Britannica* at the age of seven, which he claims to have read *in toto* (134–35). But of course this example is an aggrandizement of fiction; fiction in the Pynchonian tradition.

6. In case you were wondering, Huldreich Zwingli (1484–1531), leader of the Swiss Reformation, whom Pynchon identifies as "the man at the end of the encyclopedia" (267). Slothrop encounters "stone reminders" of Zwingli's importance in Zürich, where he has fled to avoid The Firm and research his ties to Laszlo Jamf.

7. Roger Mexico, the statistician, describes the Psi Section staff of "The White Visitation" as "wild talents—clairvoyants and mad magicians, telekinetics, astral travelers, gatherers of light . . . all the definitely 3-sigma lot" (40). Weisenburger explains that since "sigma is the designation for a standard deviation" in statistics, 3-sigma would represent a "wildly divergent 'lot' of people" (35–36).

8. Thomas Moore argues persuasively that Pynchon's erudition must be presented in an allusive manner, even in a work as encompassing and rangy as *Gravity's Rainbow:* "it is hard to see how a novel that so aspires to be 'about' a culture's entire

connectedness could afford, practically, to back every erudite bet as it goes, even if we think that such backing of bets is crucial for its artistic success. Those who now complain of Pynchon's being unreadable by virtue of his excessive formal braininess would perhaps find his novels untouchable if the novels managed somehow fully to flesh out each skeleton of allusion" (150). Moore's chapter on "The Culture of Science in *Gravity's Rainbow*" provides an excellent introduction to, and expatiation of, the scientific references in the novel.

9. In a passage that describes Franz Pökler's design work on the A3 rocket at Peenemünde in 1938, the narrator notes that the engineers study daily films of the flight tests: "There has been this strange connection between the German mind and the rapid flashing of successive stills to counterfeit movement, for at least two centuries—since Leibniz, in the process of inventing calculus, used the same approach to break up the trajectories of cannonballs through the air" (407). Implicit here, and in what follows of Pökler's sad story, is the narrator's disapproval of the reduction of nature and human lives into component parts, especially as success in the engineering problem would be measured in terms of the lives destroyed.

10. Susan Strehle discusses this passage in order to contrast modes of continuity in *Gravity's Rainbow.* Linear continuity reflects the "realist's conception of one-directional time," while molecular continuity is "woven, highly intricate, labyrinthine, and unpredictable" (53).

11. For a discussion of the destabilization of received duality and the iterative methodology in both deconstruction and chaos theory, see N. Katherine Hayles, *Chaos Bound* (175–86).

12. Economist Brian Arthur suggests that complexity is "total Taoist. In Taoism there is no inherent order. . . . The universe in Taoism is perceived as vast, amorphous, and ever-changing. . . . The elements always stay the same, yet they're always rearranging themselves" (Waldrop 330).

13. In a largely negative review essay on Pynchon's *Mason & Dixon*, James Wood complains that "Pynchon may long for the polyphonic music of the fabulist, but his novels enforce a strict binarism even as they congratulate themselves on deconstructing this binarism. Oedipa [Maas] may want 'a symmetry of choices to break down,' but Pynchon's very fiction is built on a symmetry of choices. Either utopia or dystopia; either governance or dream; either too much meaning or not enough meaning. . . . Pynchon seems to be suggesting that this ordeal of choice is what is wrong with America: it is America that forces us to choose between exile or madness" ("Levity's Rainbow" 37–38). Although several critics have insisted on the oppositional nature of Pynchon's political stance, it is my contention that his appreciation of complexity, ambiguity, and uncertainty allow him to evade the ordeal of choice.

14. Notice that the description of Imipolex G suggests the bi-stable points of behavioral conditioning: "Under suitable stimuli, the chains grow cross-links, which stiffen the molecule and increase intermolecular attraction so that this Peculiar Polymer runs far outside the known phase diagrams, from limp rubbery amorphous to amazing perfect tessellation, hardness, brilliant transparency" (699). The Pavlovian

brain, the aromatic polymer, and the floor of "The White Visitation" are all visualized as tessellated forms. The Imipolex G molecule is in the shape of "hexagons like the gold one that slides and taps above Hillary Bounce's navel" (249) and that "Pökler's old prof, Laszlo Jamf" displays "removing from his fob a gold hexagon with the German formée cross in the center, a medal of honor from IG Farben" to illustrate the tetravalency of carbon (413).

15. I am indebted to Alan J. Friedman and Manfred Puetz's discussion of "Science as Metaphor" in this early essay. But I do not think that, in considering Pynchon as "an author in search of a metaphor, a fictional scheme to ask and answer the question of what prevails in the physical and in the spiritual universe," they consider an authorial position *between* the binary positions of "order or disorder, distinction or chaos, pattern or the existential blur" (23). Only at the close of the essay do they propose that "order and chaos (and hence paranoia and antiparanoia) should not be seen as antagonists of the either/or type but as elements of one and the same universal movement" (35). But how this escape from binary antagonism might be realized in the text they do not hazard to suggest.

16. The Poisson equation, which is a power series, is given on page 140 of the novel. It is a bell-shaped curve. See Friedman and Puetz (29).

17. See Aristotle, the *Posterior Analytics*, Book I.15: "It is possible for A to be directly true of B; in the same way, it is possible for A to be directly deniable of B. By 'directly true' or 'directly deniable' I mean that there is no middle term" (180).

18. Paul Maltby discusses the appearance of "excluded middles" that perplex Oedipa Maas in *The Crying of Lot 49*. Faced with four "symmetrical" alternatives (*Crying* 171) regarding the existence of the Trystero System, Oedipa's "predicament demands that she entertain all four simultaneously, that she recognize the *coexistence of contradictory possibilities*. This, precisely, is the space of the excluded middle, of the undecidable, and it is a space in which Oedipa feels lost; for under the binary (either/or) logic sanctioned by positivism, she cannot admit the existence of this space" (*Dissident Postmodernists* 143). Waiting at last for the crying of lot 49, Oedipa "had heard all about excluded middles; they were bad shit, to be avoided; and how had it ever happened here, with the chances once so good for diversity? For it was now like walking among matrices of a great digital computer, the zeroes and ones twinned above, hanging like balanced mobiles right and left, ahead, thick, maybe endless" (*Crying* 181). Like Pointsman, Oedipa feels trapped by the binary alternatives of the Zero and the One. But the conclusion of the novel suggests that diversity might be sustained, indeterminacy preserved, by the failure to resolve the four symmetrical alternatives into a single truth. Oedipa, and Pynchon's fictions, exist in the domain of the excluded middle.

19. See Brian McHale's extended analysis in "Misreading *Gravity's Rainbow*" regarding the undecidability of narratorial address in this passage. He posits that "in the case of the Advent vespers episode we are left with a range of interpretive hypotheses for the episode's second person pronouns—narrator addressing narratee or narrator apostrophizing character or character self-address; Roger or Jessica as ad-

dressee, or Roger *and* Jessica as joint addressee—but no obvious grounds for deciding conclusively" (*Constructing Postmodernism* 105–106). Thus the indeterminacy of narrative technique reinforces the impact of the undecidable questions put by the narrator.

20. In "Trespassing Limits: Pynchon's Irony and the Law of the Excluded Middle," Francisco Collado-Rodríguez argues that Pynchon's fiction "thoroughly represents a sustained critical subversion of categorical binary thinking and, more specifically, of the traditional interpretation Western culture has given to the rigid Aristotelian Law" (472). He suggests further that twentieth-century scientific theories such as the Copenhagen interpretation of quantum mechanics and Werner Heisenberg's Principle of Indeterminacy were the century's first attempts to erode dualism. Chaos theory, Collado-Rodríguez remarks, strongly contests either/or analysis and categorical thinking, establishing a cultural link to Derridean deconstruction (476).

21. Joseph Tabbi cautions that indeterminacy in literature is not the same as the indeterminacy theorem in science, and that we should not "expect any writer, especially one so famously subversive as Pynchon, to embrace the field view unreservedly" (*Postmodern Sublime* 105). See also Friedman, who warns that "doctrinaire acceptance" of the new scientific paradigms is potentially as sterile as the ready acceptance of "non-science-related images" in *Gravity's Rainbow* ("Science and Technology" 71).

22. Friedman concludes his discussion of scientific metaphors in *Gravity's Rainbow* by restating the dualism of determinism and uncertainty: "The fundamental choices between belief in order and determinism versus a belief in disorder and lack of control are expressed in scientific images. Yet these two extremes are sterile for mankind. If determinism is absolute, no one, not even Pointsman, can be sure that he is in control and not some higher force that wants him to think so. On the other hand, the chaos of anti-paranoia denies the possibility of self-control along with all forms of determinism" ("Science and Technology" 100). Although I would not contest Friedman's statement of the binary choices, it is my contention that in regarding such extremes as "bad shit and to be avoided," Pynchon posits a mode of dwelling in the excluded middle that frees the self from the fixed, intransigent alternatives of binary logic.

23. Strehle regards Slothrop and Blicero as a schematic relationship of "antithetical doubles" much the same as my analysis of the Mexico-Pointsman pairing. She argues that "they can be imagined as zero and one, where both points represent different forms of death, and life occupies the excluded middle ground. While Slothrop abandons connections, including those linking his various selves, and thus loses human identity, Blicero pursues linear connections to their inevitable end in death and thus loses human identity. Slothrop ceases to make fictions about his own role, and Blicero constructs a perfect, closed fiction; both thereby deny themselves living roles" (50). In pursuing the antithesis between Slothrop and Blicero, however, Strehle does not suggest what sort of life form actually occupies the excluded middle ground. I would argue that Slothrop, rather than being a mere anti-Blicero, caught in a binary logic, is a force of complexity that occupies that middle space. He is alive because he is detained by no one system, but oscillates from one "part of his cycle" (*Gravity's Rainbow* 434) to another, always *between* the extremes of paranoia and anti-paranoia.

24. Although entropy appears as a leitmotif in much early Pynchon criticism, several essays are wholly devoted to the issue, especially Anne Mangel's "Maxwell's Demon, Entropy, Information: *The Crying of Lot 49*," and David Seed's "Order in Thomas Pynchon's 'Entropy.'" Tony Tanner addresses the theme generally in American fiction in "Everything Running Down" and then turns specifically to Pynchon in "Caries and Cabals" in *City of Words*. Likewise, John Kuehl in his chapter of *Alternate Worlds* on "Entropy" shows that references to entropy and increasing disorder appear frequently in the postwar American fiction of Pynchon, William Burroughs, John Barth, William Gaddis, and Ishmael Reed.

25. Weisenburger argues, "Everywhere in *Gravity's Rainbow* the parabolic arch symbolizes disease, dementia, and destruction. Its counterpart is the circular mandala, a symbol of opposites held in delicate equipoise" (10–11). But the relation of opposites is rarely delicate, most often turbulent. The attention should be given to the roiling interchange of the agents, rather than to the Zen stasis of a utopia the book never finally describes.

26. After carefully discussing entropy as a measure of disorder in a system, and the effect that a V-2 rocket can have in creating disorder, Friedman points out that life processes in *Gravity's Rainbow* appear to stand in opposition to the second law of thermodynamics. But the biological organism is not a closed system: "If it were entirely isolated, it would quickly wither and die, becoming properly disordered. For life to survive, it must consume energy, creating disorder elsewhere. . . . The rules of 'non-equilibrium' thermodynamics are more complex than the simple version of the second law. . . . *Gravity's Rainbow* is filled with images of biological order springing from disorder" ("Science and Technology" 85). In "From Entropy to Chaos-Theories," Bernd Klähn points out that Tanner and others affix "Pynchon to an apocalyptic vision of entropic decay as a general paradigm of occidental culture," but subsequent criticism argues that since Pynchon does not depict closed systems, the classical notion of entropy does not apply (420). In its place Klähn suggests that Pynchon's interest in Norbert Wiener's cybernetic concept of feedback, as "a central paradigm of a universal theory of self-organization," leads to an "evolutionary optimism" (422) that is eventually manifest in chaos theory and particularly in the work of Ilya Prigogine. For other discussions of negative entropy in Pynchon, see Mesher and Schuber.

27. In an essay that triangulates a theoretical relationship between the time-irreversible physics of Prigogine, the nomadic space of Deleuze, and the political chaos of the Zone, Martin E. Rosenberg points out that since "the self-organizing processes associated with non-equilibrium thermodynamics always refer to an open system . . . *Gravity's Rainbow* refers specifically to the 'open' state of the Zone, effaced as it is of national boundaries that were once closed frontiers" (122).

28. Cellular automata were invented by the Hungarian mathematician John von Neumann in the 1950s. They are a computer model of complex dynamical systems. A large grid of squares, or cells, are governed by rules such that they appear either black or white on the screen depending on the state of the neighboring cells. Roger Lewin points out that "Complex, dynamic patterns develop and roam across the en-

tire grid, the nature of which is influenced but not tightly determined in detail, according to the activity rules. Notice that global structure emerges from local activity rules, a characteristic of complex systems" (46–47).

29. Hanjo Berressem, in "Strangely Attractive: The Topology of Psychic and Social Space in *Vineland*," maps Edward Lorenz's strange attractor onto Jacques Lacan's psychoanalytic "schema R" on the grounds that both surfaces are möbial. I want to suggest that while Mandelbrot's fractals and Lorenz's strange attractor are the favored evidence for an order "hidden" deep within chaotic systems, the complexity theory advanced by Stuart Kauffman and Chris Langton belongs to the branch of chaotics prominently advanced by Prigogine, which emphasizes the possibility of an emergent order in systems "far from equilibrium." Thus, Langton finds the emergent order of complex adaptive systems located between the structure of the periodic attractor and the disorder of strange attractors. Pynchon's attraction to the middle zone in *Gravity's Rainbow* pursues the possibility of an emergent complexity after the collapse of wartime antitheticals and dualities.

30. For a chronology of *Gravity's Rainbow* that keys time and place, as well as providing a map of Slothrop's desultory path through North-Central Germany, see Khachig Tölölyan (33–40).

31. Daedalus typifies the ingenuity in mechanical invention for ancient Greek culture. The original labyrinth, from which Theseus escapes with the aid of Ariadne's thread, was said to have been at Cnossus on the island of Crete. The idea of the labyrinth "was perhaps suggested to later Greeks by the complex ruins of the Bronze age palace. The complicated figures of the Greek 'crane dance' were supposed to represent the convolutions of the labyrinth" (Howatson, ed., *Oxford Companion to Classical Literature* 312). The ruins of King Minos's palace were indeed complex, though the artifice of Daedalus, as a rationalized, orderly space, is complicated.

32. Weisenburger, in his own summation of the episode, remarks, "First, the episode is scattered with items from the lexicon of Teutonic myth, signified in references to ancient runes, and the Tannhäuser legend. Second, there are references to ancient Herero myth and custom. Finally these two strands—northern and southern, the white and the black—converge in a contemporary, technological mythology of rocketry" (155). It's only appropriate that the Mittelwerke episode represent the convergence of opposites.

33. In a brief note, "The Ellipsis as Architectonic in *Gravity's Rainbow*," Laurence Daw observes that the novel "contains a huge number of ellipses. Of the novel's 760 pages, only 9.7% do not contain ellipses. . . . Based on the frequency of its use in *Gravity's Rainbow*, we can speculate that the ellipsis is an architectonic for Pynchon" (54).

Chapter 8. The Superabundance of Cyberspace

1. With a flourish of hyperbole, Coover remarks, "you will often hear it said that the print medium is a doomed and outdated technology, a mere curiosity of bygone

days destined soon to be consigned forever to those dusty unattended museums we now call libraries. Indeed, the very proliferation of books and other print-based media, so prevalent in this forest-harvesting, paper-wasting age, is held to be a sign of its feverish moribundity, the last futile gasp of a once vital form before it finally passes away forever, dead as God" ("End of Books" 1).

2. Sven Birkerts defends print culture in an on-line dialog, "Page Versus Pixel," with hypertextualists Carolyn Guyer, Robert Stein, and Michael Joyce in the electronic magazine *Feed.*

3. Joseph Tabbi and Michael Wutz describe the phenomenon of technological cross-purposing as "intermediality" (citing an essay by Geoffrey Winthrop-Young and Joseph Donatelli) that suggests "a model of media interaction that takes note of the frequently catalytic interaction and intertwining of media from different historical moments" (*Reading Matters* 10) such as orality and writing, the continuation of the handwritten document in the age of print, and the simulation of the typewriter in word-processing programs.

4. Thomas S. Kuhn finds incommensurability to be the operative sign of paradigm shifts. He remarks, "Since new paradigms are born from old ones, they ordinarily incorporate much of the vocabulary and apparatus, both conceptual and manipulative, that the traditional paradigm had previously employed. But they seldom employ these borrowed elements in quite the traditional way. Within the new paradigm, old terms, concepts, and experiments fall into new relationships one with the other" (*Structure of Scientific Revolution* 149). In the present discussion of print and hypertextual paradigms, the definitive example would be the "bookmark" function of web browsers and text readers, appropriated from the old paradigm of the bound book, to "mark the place" of the reader. But like a post thrust into the shifting sand dunes of the Sahara desert, what place is marked by this virtual device in the ever-changing "docuverse" of the Internet? Who has not returned to their bookmarked "page" on the Web only to find the resource moved, the link outdated? This contingent resource locator shares a name with printed tabs, clips, and ribbons, but its function is incommensurable with that of its namesake.

5. Landow proposes that the persistence of print culture will be particularly notable "in elite and scholarly culture, and when the shift to hypertext makes it culturally dominant, it will appear so natural to the general reader-author that only specialists will notice the change or react with much nostalgia for the way things used to be" (*Hypertext 2.0* 288–89). But judging from present conditions, it seems that the scholarly elite have more readily embraced hypertext as an editing tool and writing/reading space, whereas the general reader—with little time or patience for disorienting links and balking programs—has been more hesitant to adopt digital text delivery systems.

6. Two hypertext authors, Michael Joyce and J. Yellowlees Douglas, have been installed in the new *Postmodern American Fiction: A Norton Anthology* (1998). Selections from Douglas's *I Have Said Nothing* (1993) and Joyce's *afternoon: a story* (1990) appear in the printed anthology, and are accompanied by excerpts posted on

a web site accessible to the volume's purchasers. This dual presentation, fraught with marketing and access difficulties, only serves to accentuate the incommensurable aspects of print and digital media distribution. W. W. Norton & Co. as a purveyor of textbooks will not yet presume a market for a CD-ROM anthology (though surely the capacious technology of the CD-ROM more than accommodates the bulk of text) among students; yet the presence of fiction that can only be adequately "read" in a hypertext environment militates for on-line access. At what point (in reading, or in the classroom?) does the student dispense with the printed text?

7. DeLillo provides a patently useless URL for the Miraculum web site, http://blk.www/dd.com/miraculum in *Underworld* (810). Does he intend a skeptic's irony in the use of a commercial "top-level" domain name (.com) for this site? What might be offered for sale on such a site?

8. In an interview with Terry Gross on NPR's *Fresh Air* (Thursday, 2 October 1997), DeLillo states, "I have a curious relationship to the page. I use an old manual typewriter, largely because of the sensuous sensation of hitting the keys and watching and hearing the hammers strike the page. And I have a kind of audio-visual connection to language; that is, I'm not only aware of the rhythms of my sentences, which is something I think that characterizes my work, but I'm also conscious of the physical appearance of those alphabetic units that compose a given word."

9. Barth's story appeared briefly in the on-line version of *The Atlantic Monthly* called *Atlantic Unbound*. One can imagine the difficulty encountered by visitors to the site who attempt to activate the ersatz hyperlinks in the story's text by clicking the underlined, color-coded text.

10. Barth weighs in with his own comments on the prospects of cyberfiction and virtual reality in his essay, "Virtuality." He acknowledges both a generational and methodological transition from print to digital culture, suggesting nonetheless that the traditional novelist will not soon be run out of the business: "Although few of us still prefer to compose our sentences in longhand before turning them into pixels on a VDT en route to their returning into print on a page, and a few more still prefer to bang away at an old-fashioned typewriter and eschew computers altogether, the superconvenient word-processor has become, in only a dozen years, the production mode of choice for most writers of most kinds of writing, whether or not it affects the quality of the product. Although interactive computer-fiction (especially as it comes to include whole repertories of graphic, cinematic, and auditory effects) is too fascinating not to become yet another competitor for audience attention, one doubts that it will have nearly the market-share effect on 'straight' fiction-reading that movies and television have had. Those who still read literature for pleasure at all (no more than 10 percent of the adult U.S. population, says the *New York Times*) are likely to go on preferring, most of the time, the customary division of labor between Teller and Told" (n.p.).

11. George P. Landow grants in theory the association of hypertext networks and the rhizome: "when Deleuze and Guattari write that a rhizome 'has neither beginning nor end, but always a middle (milieu) from which it grows and which it overspills,'

they describe something that has much in common with the kind of quasi-anarchic networked hypertext one encounters in the World Wide Web, but when in their next sentence they add that the rhizome 'is composed not of units but of dimensions, or rather directions in motion' (21), the parallel seems harder to complete. The rhizome is essentially a counter paradigm, not something realizable in any time or culture; but it can serve as an ideal for hypertext" (*Hypertext 2.0* 42).

12. In 1991 the World Wide Web was created as a component of the Internet by Tim Berners-Lee at the European Laboratory for Particle Physics (CERN) as an Internet information distribution system to be shared among research groups. Although some elements of the Internet such as ftp, gopher, WAIS, and newsgroups may be controlled by an institution or owned by an individual, the WWW has been open to public development, without an imposed structure, although the HyperText Transmission Protocol (HTTP) is administered by the W3 Consortium. The state of the Web at any given moment is very much described by Deleuze and Guattari's proposition of a "global result synchronized without a central agency."

13. William Safire points out, in "Return of the Luddites," that between 1811 and 1816 the Luddites "waged a war against the serflike conditions spawned by the users of textile machinery. 'If workmen dislike certain machines,' explained The Nottingham Review in 1811, 'it was because of the use to which they were put, not because they were machines or because they were new'" (34). The Luddites were originally protesting for decent living and working conditions, not because they were opposed to technological progress. For a fictional revision of socialist-Luddite uprising set in London in 1855, see William Gibson and Bruce Sterling, *The Difference Engine* (1991).

14. Henry Ford's attempts to "curry favor with labor" in 1914 are recounted in Todd Alden, "Here Nothing, Always Nothing: A New York Dada Index, etc.," in *Making Mischief: Dada Invades New York*, ed. Francis M. Naumann (New York: Whitney Museum of American Art/ Harry N. Abrams, 1996), p. 42.

15. Since the publication of Sales's article, the London tabloids (notorious for their intrusions into the private lives of the royal family and the filthy rich; even more so since the death of Princess Diana) published a paparazzo photograph of Pynchon and son. James Bone's photograph and accompanying article appeared in *The Times* of London on 14 June 1997; a commentary on the exposé by Bruce Gulp appeared in *The Globe and Mail*, New York edition, on Saturday, 26 July 1997. According to Gulp, Bone had in fact pursued the lead "provided by *New York Magazine*. It ran a short item saying that Mr. Pynchon was living in the Upper West Side (though no address was given). Through Lexis-Nexis, an on-line service that sells personal public information to credit agencies and newspapers, Mr. Bone found his man—a Thomas Ruggles Pynchon Jr."

16. See Michael Seltzer, *Terrorist Chic*.

17. In "The Precession of Simulacra," Jean Baudrillard relates the "indifferentiation" of the real and the simulation to Marshall McLuhan's adage that "the medium is the message": "The medium is no longer identifiable as such, and the confusion of

the medium and the message (McLuhan) is the first great formula of this new era. There is no longer a medium in the literal sense: it is now intangible, diffused, and diffracted in the real, and one can no longer even say that the medium is altered by it" (*Simulacra and Simulation* 30). This condition, what I would call the immanence of information, describes the state of the informational matrix as it is transformed in Gibson's novel.

18. In "Simulacra and Science Fiction," Baudrillard remarks that at "the end of science fiction—the era of hyperreality begins" (*Simulacra and Simulation* 124). Gibson eschews what Baudrillard finds in science fiction: the "projection of the real world of [industrial] production" into the future without presenting a qualitatively different schema; or, in Gibson's terms, an "extrapolation." Baudrillard's preference for a projection that is "totally reabsorbed in the implosive era of models . . . that of simulation in the cybernetic sense" (*Simulacra and Simulation* 122) is closely related to Gibson's conception of cyberspace. Baudrillard's example of cybernetic fiction is J. G. Ballard's *Crash* (1973).

19. An extension of the "simstim" experience has appeared in *Star Trek: The Next Generation* as the "holodeck," in which holographic images are projected with apparently life-like interaction with actual crew members. In *Neuromancer,* Case and his associates attend an entertainment in a restaurant on the orbital Freeside station featuring "the holographic cabaret of Mr. Peter Riviera" (138). See also Baudrillard's essay on "Holograms" in *Simulacra and Simulation*.

20. In *What Is Post-Modernism?*, Jencks argues that the "most visible shift in the post-modern world is towards pluralism and cultural eclecticism" (50).

21. Responding to McCaffery's suggestion that the "breakdown of distinctions—between pop culture and 'serious' culture, different genres, different art forms—seems to have a liberating effect," Gibson replies, "The idea that all this stuff is potentially grist for your mill has been very liberating. This process of cultural mongrelization seems to be what postmodernism is all about. The result is a generation of people (some of whom are artists) whose tastes are wildly eclectic—people who are hip to punk music and Mozart, who rent these terrible horror and SF videos from the 7–11 one night and then invite you to a mud wrestling match or a poetry reading the next. . . . I know I don't have a sense of writing as being divided up into different *compartments*, and I don't separate literature from the other arts. Fiction, television, music, film—all provide material in the form of images and phrases and codes that creep into my writing in ways both deliberate and unconscious" ("Interview" 266).

22. McHale includes as Appendix 10.1 and 10.2 to his chapter on "POSTcyberMODERNpunkISM" in *Constructing Postmodernism* the relevant passages from Gibson's *Neuromancer* and Acker's *Empire of the Senseless* (239–42). McHale's essay is included with modification in McCaffery, ed., *Storming the Reality Studio* (308–23).

23. For a somewhat broader definition of encyclopædic narrative, see Edward Mendelson, "Gravity's Encyclopedia," in Levine and Leverenz, eds., *Mindful Pleasures* (161–95).

24. In an interview with Michael Tortorello, Powers responds to the question of

information overload: "It's not clear whether the stockpile is growing faster than the index. There's almost that ironic sense of feeling on Monday/Wednesday/Friday that we've been to the computer just in time to facilitate the management of the amount of data that we're producing; and then on Tuesday/Thursday/and the weekends thinking it's actually a kind of after-the-fact finger in the dike, too little, too late. I suppose the question is not necessarily whether the index is staying on top of the stockpile, but whether any human being can even keep up with the index. I do talk about that to some degree in *Galatea* with the little discussion between Richard and Helen. He describes the paradox of the archive, which is, once you have a permanent medium of representation and recording, the notion of individual life gets lost in the notion of a constantly accreting history. And we're now at the point where more books get written in a year than an individual could read in a lifetime. So this whole sense of living in a historic continuity with existing material breaks down under its own weight, collapses from its own magnitude. And Helen says 'books will have to die.' And Richard says, we've tried that, and they're called magazines. And in a way, there's a real truth to that. Books sort of are becoming magazines. They're marked with that date upon which they're supposed to be removed from the racks. In this sense, a continuing conversation—ongoing, all the way back to the beginning—is hard to sustain" ("Industrial Evolution" n.p.).

25. As a counterpoint to the present era of media saturation, the narrator points out that—on the day that defines for DeLillo the beginning of the postwar era—there's "a man on 12th Street in Brooklyn who has attached a tape machine to his radio so he can record the voice of Russ Hodges broadcasting the game. The man doesn't know why he's doing this. It is just an impulse, a fancy, it is like hearing the game twice, it is like being young and being old, and this will turn out to be the only known recording of Russ' famous account of the final moments of the game" (*Underworld* 32).

26. Although it may be an instance of novelistic coincidence, DeLillo's character "Cotter Martin" (who is black) may be based on an historical figure, Robert Cotter (who is white), who was arrested for retrieving a baseball hit into the bleachers of Baker Bowl during a Philadelphia Phillies game in 1922. At that time, fans were expected to return baseballs as the property of the home team, and Cotter's act cost him a day in jail. Adverse publicity, and Cotter's subsequent release by a sympathetic judge, worked to change this policy. DeLillo's Cotter Martin has to wrestle the home-run ball from another determined fan, a scene repeated frequently in the race between Mark McGwire and Sammy Sosa for Roger Maris's single-season home run record.

27. As opposed to pixilated, "behaving as if mentally imbalanced; very eccentric. Whimsical, prankish. From pixy, a fairylike or elfin creature, especially one that is playful or mischievous." *American Heritage Dictionary.*

Works Cited

Aarseth, Espen J. *Cybertext: Perspectives on Ergodic Literature.* Baltimore and London: Johns Hopkins University Press, 1997.

Abish, Walter. *Alphabetical Africa.* New York: New Directions, 1974.

Acker, Kathy. *Algeria: A Series of Invocations Because Nothing Else Works.* London: Aloes Books, 1984.

———. *Blood and Guts in High School.* New York: Grove, 1978.

———. *Bodies of Work: Essays by Kathy Acker.* London and New York: Serpent's Tail, 1997.

———. "A Conversation with Kathy Acker." With Ellen G. Friedman. *Review of Contemporary Fiction* 9.3 (1989): 12–22.

———. "Devoured by Myths: An Interview with Sylvère Lotringer." Acker, *Hannibal Lecter* 1–24.

———. *Empire of the Senseless.* New York: Grove, 1988.

———. "A Few Notes on Two of My Books." *Review of Contemporary Fiction* 9.3 (1989): 31–36.

———. *Hannibal Lecter, My Father.* Ed. Sylvère Lotringer. New York: Semiotext(e), 1991.

———. *In Memoriam to Identity.* New York: Grove, 1990.

———. *Literal Madness: Kathy Goes to Haiti; My Death My Life By Pier Paolo Pasolini; Florida.* New York: Grove, 1988.

———. "The Path of Abjection: An Interview with Kathy Acker." With Larry McCaffery. Larry McCaffery, ed. *Some Other Frequency: Interviews with Innovative American Authors.* Philadelphia: University of Pennsylvania Press, 1996. 14–35.

Adams, Henry. *The Education of Henry Adams.* Boston: Houghton Mifflin, 1961.

Alden, Todd. "Here Nothing, Always Nothing: A New York Dada Index, etc." *Making Mischief: Dada Invades New York.* Ed. Francis M. Naumann. New York: Whitney Museum of American Art/Harry N. Abrams, 1996. 33–175.

Ammons, A. R. *Garbage.* New York and London: Norton, 1993.

Andersen, Richard. *Robert Coover.* Boston: Twayne, 1981.

Anderson, Perry. *The Origins of Postmodernity.* New York and London: Verso, 1998.

Argyros, Alexander J. *A Blessed Rage for Order: Deconstruction, Evolution, and Chaos.* Ann Arbor: University of Michigan Press, 1991.
Aristotle. "Posterior Analytics, Book I." *The Philosophy of Aristotle.* Ed. Renford Bambrough. Trans. A. E. Wardman. New York: Mentor-NAL, 1963. 160–204.
Arnheim, Rudolf. *Entropy and Art: An Essay on Disorder and Order.* Berkeley and Los Angeles: University of California Press, 1971.
Ashbery, John. "The System." *Three Poems.* New York: Penguin, 1977.
Auster, Paul. *The Art of Hunger: Essays, Prefaces, Interviews and* The Red Notebook. New York: Penguin, 1993.
———. "Interview with Larry McCaffery and Sinda Gregory." *The Art of Hunger.* 277–320.
———. *The Music of Chance.* New York: Viking, 1990.
Ballard, J. G. *Crash.* New York: Farrar, Straus & Giroux, 1973.
Barrow, John D. *The Artful Universe.* Oxford, Eng.: Clarendon Press, 1995.
Barth, John. "Click." *Atlantic Monthly* December 1997: 81–96.
———. *The Floating Opera and The End of the Road.* 1956, 1958. New York: Anchor Books, 1988.
———. *Further Fridays: Essays, Lectures, and Other Nonfiction, 1984–94.* Boston: Little, Brown, 1995.
———. *The Last Voyage of Somebody the Sailor.* Boston: Little, Brown, 1991.
———. *LETTERS.* 1979. Normal, IL: Dalkey Archive Press, 1994.
———. "The Literature of Exhaustion." *The Friday Book: Essays and Other Nonfiction.* New York: Putnam, 1984. 62–76.
———. *Lost in the Funhouse: Fiction for Print, Tape, Live Voice.* New York: Bantam, 1969.
———. "Virtuality." *Johns Hopkins Magazine* (Sept. 1994): 10 pars. 30 Jun. 1998 <http://www.jhu.edu/~jhumag/994web/culture1.html>.
Barthelme, Donald. *Snow White.* New York: Atheneum, 1967.
Barthes, Roland. *Critical Essays.* Trans. Richard Howard. Evanston, IL: Northwestern University Press, 1972.
———. *Elements of Semiology.* Trans. Annette Lavers and Colin Smith. New York: Hill and Wang, 1968.
Baudrillard, Jean. *Simulacra and Simulation.* Trans. Sheila Faria Glaser. Ann Arbor: University of Michigan Press, 1994.
Baxter, Charles. "In the Suicide Seat: Reading John Hawkes's *Travesty.*" *Georgia Review* 34 (1980): 871–85.
Beckett, Samuel. *Three Novels: Molloy; Malone Dies; The Unnamable.* New York: Grove Press, 1958.
———. *Watt.* New York: Grove Press, 1953.
Bellos, David. "Georges Perec's Puzzling Style." *Scripsi* 5.1 (1988): 63–78.
Bénabou, Marcel. "Rule and Constraint." Motte 40–47.
Berman, Neil. "Coover's *Universal Baseball Association:* Play as Personalized Myth." *Modern Fiction Studies* 24 (1978): 209–222.

Berressem, Hanjo. "Strangely Attractive: The Topology of Psychic and Social Space in *Vineland*." *Pynchon Notes* 34–35 (Spring-Fall 1994): 38–55.

Berryman, Charles. "Hawkes and Poe: *Travesty*." *Modern Fiction Studies* 29 (1983): 643–54.

Bertens, Hans. *The Idea of the Postmodern: A History*. New York: Routledge, 1995.

Birkerts, Sven. *The Gutenberg Elegies: The Fate of Reading in an Electronic Age*. Boston and London: Faber and Faber, 1994.

Birkerts, Sven, Carolyn Guyer, Michael Joyce, and Robert Stein. "Page Versus Pixel." *Feed* (1995): 26 pars. 10 Jan. 1999 <http://www.feedmag.com/95.05dialog1.html>.

Bloom, Harold. "Introduction." Bloom, *Thomas Pynchon's "Gravity's Rainbow"* 1–9.

———, ed. *Thomas Pynchon*. New York: Chelsea House, 1986.

———, ed. *Thomas Pynchon's "Gravity's Rainbow."* New York: Chelsea House, 1986.

Borges, Jorge Luis. *Labyrinths: Selected Stories and Other Writings*. Ed. Donald A. Yates and James E. Irby. New York: New Directions, 1964.

Bradbury, Malcolm, and James McFarlane, eds. *Modernism, 1890–1930*. New York: Penguin, 1976.

Brooke-Rose, Christine. *Amalgamemnon*. Manchester, Eng.: Carcanet, 1984.

Bruce, Donald, and Terry Butler. "Towards the Discourse of the Commune: Characteristic Phenomena in Jules Vallès's *Jacques Vingtras*." *Texte* 13–14 (1993): 219–49.

Burroughs, William. *The Ticket that Exploded*. New York: Grove Press, 1968.

Butler, Christopher. *After the Wake: The Contemporary Avant-Garde*. Oxford: Oxford University Press, 1980.

Cage, John. *Fontana Mix & Solo for Voice 2*. Perf. Eberhard Blum. Hat Hut, 1993.

———. *Silence*. Middletown, CT: Wesleyan University Press, 1961.

Caldwell, Roy C., Jr. "Of Hobby-Horses, Baseball, and Narrative: Coover's *Universal Baseball Association*." *Modern Fiction Studies* 33 (1987): 161–71.

Calvino, Italo. *The Castle of Crossed Destinies*. Trans. William Weaver. New York: Harcourt Brace Jovanovich, 1976.

Campbell, Jeremy. *Grammatical Man: Information, Entropy, Language, and Life*. New York: Simon and Schuster, 1982.

Civello, Paul. *American Literary Naturalism and its Twentieth Century Transformations: Frank Norris, Ernest Hemingway, Don DeLillo*. Athens and London: University of Georgia Press, 1994.

Clemens, Samuel Langhorne. *Adventures of Huckleberry Finn*. Ed. Sculley Bradley, Richmond Croom Beatty, and E. Hudson Long. New York: Norton, 1962.

Clerc, Charles, ed. *Approaches to "Gravity's Rainbow."* Columbus: Ohio State University Press, 1983.

Collado-Rodríguez, Francisco. "Trespassing Limits: Pynchon's Irony and the Law of the Excluded Middle." *Oklahoma City University Law Review* 24 (1999): 471–503.

Conte, Joseph. "The Uncertain Predictor: Calvino's *Castle* of Tarot Cards." *Literature and Science*. Ed. Donald Bruce and Anthony Purdy. Amsterdam and Atlanta, GA: Editions Rodopi, 1994. 131–47.

Coover, Robert. "The End of Books." *New York Times Book Review* 21 Jun. 1992: 1+.

——. "Hyperfiction: Novels for the Computer." *New York Times Book Review* 29 Aug. 1993: 1+.

——. "An Interview with Robert Coover." With Larry McCaffery. LeClair and McCaffery, *Anything Can Happen*. 63–78.

——. *Pricksongs and Descants*. New York: Plume-NAL, 1969.

——. "Robert Coover." Interview with Frank Gado. *First Person: Conversations on Writers and Writing*. Ed. Frank Gado. Schenectady, NY: Union College Press, 1973. 142–59.

——. *The Universal Baseball Association, Inc., J. Henry Waugh, Prop.* New York: Plume-NAL, 1968.

Cope, Jackson I. *Robert Coover's Fictions*. Baltimore and London: Johns Hopkins University Press, 1986.

Cowley, Julian. "Gilbert Sorrentino." *Dictionary of Literary Biography* 173: *American Novelists Since World War II*, Fifth Series. Ed. James R. Giles and Wanda H. Giles. Detroit: Gale Research Press, 1996. 249–59.

Creeley, Robert. *The Company*. Providence: Burning Deck, 1988.

Daw, Laurence. "The Ellipsis as Architectonic in *Gravity's Rainbow*." *Pynchon Notes* 11 (February 1983): 54–56.

Deleuze, Gilles, and Félix Guattari. *A Thousand Plateaus: Capitalism and Schizophrenia*. Trans. Brian Massumi. Minneapolis and London: University of Minnesota Press, 1987.

DeLillo, Don. "The Art of Fiction CXXXV." Interview with Adam Begley. *Paris Review* 128 (Fall 1993): 274–306.

——. *Great Jones Street*. Boston: Houghton Mifflin, 1973.

——. Interview with Ray Suarez. National Public Radio Book Club of the Air (4 Aug. 1994): 8 par. 5 Oct. 1998. <http://haas.berkeley.edu/~gardner/technoise.htm>.

——. "An Interview with Don DeLillo." Interview with Tom LeClair. LeClair and McCaffery, *Anything Can Happen*. 79–90.

——. *Libra*. New York: Viking, 1988.

——. "'An Outsider in This Society': An Interview with Don DeLillo." Interview with Anthony DeCurtis. *Introducing Don DeLillo*. Ed. Frank Lentricchia. Durham and London: Duke University Press, 1991. 43–66.

——. *Running Dog*. New York: Knopf, 1978.

——. *Underworld*. New York: Scribner, 1997.

——. *White Noise*. New York: Viking, 1985.

Dick, Leslie. "Feminism, Writing, Postmodernism." *From My Guy to Sci-Fi: Genre and Women's Writing in the Postmodern World*. Ed. Helen Carr. London: Pandora Press, 1989. 204–14.

Dickinson, Emily. *Final Harvest: Emily Dickinson's Poems*. Ed. Thomas H. Johnson. Boston: Little, Brown, 1961.

Dix, Douglas Shields. "Kathy Acker's *Don Quixote*: Nomad Writing." *Review of Contemporary Fiction* 9.3 (1989): 56–62.

Donoghue, Denis. *Connoisseurs of Chaos: Ideas of Order in Modern American Poetry*, 2nd ed. New York: Columbia University Press, 1984.

Duncan, Robert. *The Opening of the Field*. New York: New Directions, 1960.

Eco, Umberto. *The Aesthetics of Chaosmos: The Middle Ages of James Joyce*. Trans. Ellen Esrock. Cambridge, MA: Harvard University Press, 1989.

——. *The Open Work*. Trans. Anna Cancogni. Cambridge, MA: Harvard University Press, 1989.

Ekeland, Ivar. *The Broken Dice and Other Mathematical Tales of Chance*. Trans. Carol Volk. Chicago and London: University of Chicago Press, 1993.

Eliot, T. S. *The Complete Poems and Plays, 1909–1950*. New York: Harcourt, Brace & World, 1971.

——. "*Ulysses*, Order, and Myth." *Selected Prose of T. S. Eliot*. Ed. Frank Kermode. New York: Harcourt Brace Jovanovich, 1975. 175–78.

Federman, Raymond. *The Twofold Vibration*. Bloomington: Indiana University Press, 1982.

Fletcher, John, and Malcolm Bradbury. "The Introverted Novel." Bradbury and McFarlane, eds. *Modernism, 1890–1930*. 394–415.

Foucault, Michel. *The Archaeology of Knowledge and the Discourse on Language*. Trans. A. M. Sheridan Smith. New York: Pantheon, 1972.

——. *Discipline and Punish: The Birth of the Prison*. Trans. Alan Sheridan. New York: Vintage, 1979.

——. *The History of Sexuality*. Volume I: An Introduction. Trans. Robert Hurley. New York: Random House, 1978.

Friedman, Alan J. "Science and Technology." Clerc 69–102.

Friedman, Alan J., and Manfred Puetz. "*Gravity's Rainbow*: Science as Metaphor." Bloom, *Thomas Pynchon* 23–35.

Friedman, Ellen G. "'Now Eat Your Mind': An Introduction to the Works of Kathy Acker." *Review of Contemporary Fiction* 9.3 (1989): 37–49.

Gaddis, William. *A Frolic of His Own*. New York: Scribner, 1994.

Geyh, Paula, Fred G. Leebron, and Andrew Levy, eds. *Postmodern American Fiction: A Norton Anthology*. New York: W. W. Norton, 1998.

Gibson, William. "An Interview with William Gibson." Interview with Larry McCaffery. McCaffery, *Storming the Reality Studio*. 263–85.

——. *Mona Lisa Overdrive*. New York: Bantam, 1988.

——. *Neuromancer*. New York: Ace, 1984.

Gibson, William, and Bruce Sterling. *The Difference Engine*. New York: Bantam, 1991.

Gleick, James. *Chaos: Making a New Science*. New York: Viking, 1987.

Gordon, Lois. *Robert Coover: The Universal Fictionmaking Process*. Carbondale and Evansville: Southern Illinois University Press, 1983.

Hansen, Arlen J. "The Dice of God: Einstein, Heisenberg, and Robert Coover." *Novel* 10.1 (1976): 49–58.

Haraway, Donna. *Simians, Cyborgs and Women: The Reinvention of Nature*. New York: Routledge, 1991.

Harris, Paul A. "The Invention of Forms: Perec's *Life, A User's Manual* and a Virtual Sense of the Real." *SubStance* 23.2 (1994): 56–85.

Harris, Robert R. "A Talk with Don DeLillo." *New York Times Book Review,* 10 Oct. 1982: 26.

Hassan, Ihab. *The Postmodern Turn: Essays in Postmodern Theory and Culture.* Columbus: Ohio State University Press, 1987.

Hawkes, John. "A Conversation with John Hawkes." Interview with Paul Emmett and Richard Vine. *Chicago Review* 28 (1976): 163–71.

———. *Travesty.* New York: New Directions, 1976.

Hayles, N. Katherine. "Chance Operations: Cagean Paradox and Contemporary Science." Perloff and Junkerman 226–41.

———, ed. *Chaos and Order: Complex Dynamics in Literature and Science.* Chicago and London: University of Chicago Press, 1991.

———. *Chaos Bound: Orderly Disorder in Contemporary Literature and Science.* Ithaca and London: Cornell University Press, 1990.

———. *The Cosmic Web: Scientific Field Models and Literary Strategies in the Twentieth Century.* Ithaca: Cornell University Press, 1984.

———. "Postmodern Parataxis: Embodied Texts, Weightless Information." *American Literary History* 2 (Fall 1990): 394–421.

Henderson, Eric Paul. "Structured Visions in the Novels of John Hawkes." *DAI* 46 (1986): 3034A. The University of Western Ontario (Canada), 1985.

Hofstadter, Douglas. *Gödel, Escher, Bach: An Eternal Golden Braid.* New York: Basic Books, 1979.

Horgan, John. *The End of Science: Facing the Limits of Knowledge in the Twilight of the Scientific Age.* Reading, MA: Addison-Wesley, 1996.

Howatson, M. C., ed. *The Oxford Companion to Classical Literature.* 2nd ed. New York and Oxford: Oxford University Press, 1989.

Hutcheon, Linda. *The Politics of Postmodernism.* New York: Routledge, 1989.

Jacobs, Naomi. "Kathy Acker and the Plagiarized Self." *Review of Contemporary Fiction* 9.3 (1989): 50–55.

Jakobson, Roman. *Language in Literature.* Ed. Krystyna Pomorska and Stephen Rudy. Cambridge, MA and London: Harvard University Press, 1987.

James, C. L. R. *The Black Jacobins: Toussaint L'Ouverture and the San Domingo Revolution.* 2nd ed., rev. New York: Vintage, 1963

Jameson, Fredric. *Postmodernism, or, The Cultural Logic of Late Capitalism.* Durham, NC: Duke University Press, 1991.

Jencks, Charles. *Post-Modernism: The New Classicism in Art and Architecture.* New York: Rizzoli, 1987.

———. *What Is Post-Modernism?* 4th ed. London: Academy Editions, 1996.

Johnson, B. S. *Christie Malry's Own Double-Entry.* New York: New Directions, 1973.

———. "Introduction to *Aren't You Rather Young to Be Writing Your Memoirs?*" *Review of Contemporary Fiction* 5.2 (1985): 4–13.

Johnston, John. *Information Multiplicity: American Fiction in the Age of Media Saturation*. Baltimore and London: Johns Hopkins University Press, 1998.

Joyce, James. *Finnegans Wake*. New York: Penguin, 1976.

——. *A Portrait of the Artist as a Young Man*. The Viking Critical Library. Ed. Chester G. Anderson. New York: Viking, 1968.

Joyce, Michael. *Of Two Minds: Hypertext Pedagogy and Poetics*. Ann Arbor: University of Michigan Press, 1995.

Kauffman, Stuart. *At Home in the Universe: The Search for the Laws of Self-Organization and Complexity*. New York and Oxford: Oxford University Press, 1995.

Keesey, Douglas. *Don DeLillo*. New York: Twayne, 1993.

Klähn, Bernd. "From Entropy to Chaos-Theories: Thermodynamic Models of Historical Evolution in the Novels of Thomas Pynchon and Robert Coover." *Reconstructing American Literary and Historical Studies*. Eds. Günter H. Lenz, Hartmut Keil, and Sabine Bröck-Sallah. New York: St. Martin's Press, 1990. 418–31.

Korn, Henri. "Society and the Scientific Imagination." The 1998 Capen Lecture in the Humanities. SUNY at Buffalo. 20 Oct. 1998.

Kuehl, John. *Alternate Worlds: A Study of Postmodern Antirealistic American Fiction*. New York and London: New York University Press, 1989.

Kuhn, Thomas S. *The Structure of Scientific Revolutions*. 3rd ed. Chicago and London: University of Chicago Press, 1996.

Landow, George P. *Hypertext 2.0: The Convergence of Contemporary Critical Theory and Technology*. Baltimore and London: Johns Hopkins University Press, 1997.

——. "What's a Critic to Do?: Critical Theory in the Age of Hypertext." *Hyper/Text/Theory*. Ed. Landow. Baltimore and London: Johns Hopkins University Press, 1994.

Landow, George P., and Paul Delany. "Hypertext, Hypermedia and Literary Studies: The State of the Art." *Hypermedia and Literary Studies*. Ed. Landow and Delany. Cambridge, MA and London: MIT Press, 1991. 3–50.

LeClair, Thomas. *The Art of Excess: Mastery in Contemporary American Fiction*. Urbana and Chicago: University of Illinois Press, 1989.

——. *In the Loop: Don DeLillo and the Systems Novel*. Urbana and Chicago: University of Illinois Press, 1987.

——. "The Novelists: John Hawkes." *New Republic* 10 Nov. 1979: 26–29.

——. "The Prodigious Fiction of Richard Powers, William Vollman, and David Foster Wallace." *Critique: Studies in Contemporary Fiction* 38 (1996): 12–37.

LeClair, Thomas, and Larry McCaffery, eds. *Anything Can Happen: Interviews with Contemporary American Novelists*. Urbana: University of Illinois Press, 1983.

Leonard, W. E., and S. B. Smith, eds. *T. Lucreti Cari De Rerum Natura, libri sex*. Madison: University of Wisconsin Press, 1942.

Levine, George, and David Leverenz, eds. *Mindful Pleasures: Essays on Thomas Pynchon*. Boston: Little, Brown, 1976.

Lewin, Roger. *Complexity: Life at the Edge of Chaos*. New York: Macmillan, 1992.

Livingston, Paisley, ed. *Disorder and Order: Proceedings of the Stanford International Symposium (Sept. 14–16, 1981)*. Saratoga: Anma Libri, 1984.

Lucretius. *On Nature*. Trans. Russell M. Geer. Indianapolis, IN: Bobbs-Merrill, 1965.

Lyotard, Jean-François. *The Postmodern Condition: A Report on Knowledge*. Trans. Geoff Bennington and Brian Massumi. Minneapolis: University of Minnesota Press, 1984.

Mackey, Louis. *Fact, Fiction, and Representation: Four Novels by Gilbert Sorrentino*. Columbia, SC: Camden House, 1996.

Magné, Bernard. "Transformations of Constraint." Trans. David Bellos. *Review of Contemporary Fiction* 13.1 (1993): 111–23.

Maltby, Paul. *Dissident Postmodernists: Barthelme, Coover, Pynchon*. Philadelphia: University of Pennsylvania Press, 1991.

——. "The Romantic Metaphysics of Don DeLillo." *Contemporary Literature* 37 (1996): 258–77.

Mandelbrot, Benoit. *The Fractal Geometry of Nature*. New York: W. H. Freeman, 1983.

Mangel, Anne. "Maxwell's Demon, Entropy, Information: *The Crying of Lot 49*." Levine and Leverenz 87–100.

Mathews, Harry. *Cigarettes*. New York: Wiedenfeld & Nicolson, 1987.

——. "A Conversation with Harry Mathews." Interview with John Ash. *Review of Contemporary Fiction* 7.3 (1987): 21–32.

——. "Georges Perec." *Grand Street* 3.1 (1983): 136–45.

——. "An Interview with Harry Mathews." With Lytle Shaw. *Chicago Review* 43.2 (1997): 36–52.

——. "John Ashbery Interviewing Harry Mathews." With John Ashbery. *Review of Contemporary Fiction* 7.3 (1987): 36–48.

——. *The Journalist*. Normal, IL: Dalkey Archive Press, 1997.

——. "Mathews's Algorithm." Motte 126–39.

——. "Oulipo." *Word Ways: The Journal of Recreational Linguistics* 9.2 (1976): 67–74.

——. "Vanishing Point." *The Avant-Garde Tradition in Literature*. Ed. Richard Kostelanetz. Buffalo, NY: Prometheus Books, 1982. 310–14.

Mathews, Harry, and Alastair Brotchie, eds. *Oulipo Compendium*. London: Atlas Press, 1998.

McAlmon, Robert. *Being Geniuses Together, 1920–1930*. Rev. ed. Garden City, NY: Doubleday, 1968.

McCaffery, Larry. "Robert Coover." *Dictionary of Literary Biography 2: American Novelists Since World War II*. Eds. Jeffrey Helterman and Richard Layman. Detroit: Gale Research Press, 1978. 106–121.

——, ed. *Storming the Reality Studio: A Casebook of Cyberpunk and Postmodern Science Fiction*. Durham, NC and London: Duke University Press, 1991.

McHale, Brian. *Constructing Postmodernism*. New York and London: Routledge, 1992.

——. *Postmodernist Fiction*. New York and London: Methuen, 1987.

McPheron, William. *Gilbert Sorrentino: A Descriptive Bibliography*. Elmwood Park, IL: Dalkey Archive Press, 1991.

Melville, Herman. *Novels and Tales, Volume III*. New York: Library of America, 1984.

Mendelson, Edward. "Gravity's Encyclopedia." Levine and Leverenz 161–95.

Mesher, David. "Negative Entropy and the Form of *Gravity's Rainbow*." *Research Studies* 49 (1981): 162–70.

Miller, Walter M., Jr. *A Canticle for Leibowitz*. New York: Bantam, 1972.

Moles, Abraham. *Information Theory and Esthetic Perception*. Trans. Joel E. Cohen. Urbana and London: University of Illinois Press, 1966. Trans. of *Théorie de l'information et perception esthétique*. 1958.

Moore, Thomas. *The Style of Connectedness: "Gravity's Rainbow" and Thomas Pynchon*. Columbia: University of Missouri Press, 1987.

Motte, Warren F., Jr., ed. *Oulipo: A Primer of Potential Literature*. Lincoln and London: University of Nebraska Press, 1986.

Moulthrop, Stuart. *Victory Garden*. Cambridge, MA: Eastgate Systems, 1991.

Nadel, Ira B. "'A Precision of Appeal': Louis Zukofsky and the *Index of American Design*." *Upper Limit Music: The Writing of Louis Zukofsky*. Ed. Mark Scroggins. Tuscaloosa and London: University of Alabama Press, 1997. 112–26.

Natoli, Joseph. *Mots d'Ordre: Disorder in Literary Worlds*. Albany: State University of New York Press, 1992.

Nicolis, Grégoire, and Ilya Prigogine. *Self-Organization in Non-Equilibrium Systems: From Dissipative Structures to Order Through Fluctuations*. New York: Wiley, 1977.

North, Michael. *The Political Aesthetic of Yeats, Eliot, and Pound*. New York: Cambridge University Press, 1991.

O'Brien, Flann. *At Swim-Two-Birds*. New York: NAL, 1976.

O'Donnell, Patrick. *John Hawkes*. Boston: Twayne, 1982.

Paulson, William R. *The Noise of Culture: Literary Texts in a World of Information*. Ithaca: Cornell University Press, 1988.

Perec, Georges. *A Void*. Trans. Gilbert Adair. London: Harvill, 1994. Trans. of *La Disparition*. Paris: Denoël, 1969.

——. *Life, A User's Manual*. Trans. David Bellos. Boston: David R. Godine, 1987. Trans. of *La Vie, mode d'emploi*. Paris: Hachette, 1978.

Perelman, Bob. *The Trouble with Genius: Reading Pound, Joyce, Stein, and Zukofsky*. Berkeley and Los Angeles: University of California Press, 1994.

Perloff, Marjorie. "The Return of the (Numerical) Repressed: From Free Verse to Procedural Play." *Radical Artifice: Writing Poetry in the Age of Media*. Chicago and London: University of Chicago Press, 1991. 134–70.

Perloff, Marjorie, and Charles Junkerman, eds. *John Cage: Composed in America*. Chicago and London: University of Chicago Press, 1994.

Peters, Greg Lewis. "Dominance and Subversion in Kathy Acker." *State of the Fantastic: Studies in the Theory and Practice of Fantastic Literature and Film*. Ed. Nicholas Ruddick. Westport, CT: Greenwood Press, 1990. 149–56.

Porush, David. "Fictions as Dissipative Structures: Prigogine's Theory and Postmodernism's Roadshow." Hayles, *Chaos and Order* 54–84.

———. "Making Chaos: Two Views of a New Science." *The Literature of Science: Perspectives on Popular Scientific Writing.* Ed. Murdo William McRae. Athens and London: University of Georgia Press, 1993. 152–68.

Pound, Ezra. *The Cantos of Ezra Pound.* New York: New Directions, 1989.

Powers, Richard. *Galatea 2.2.* New York: Farrar, Straus & Giroux, 1995.

———. *The Gold Bug Variations.* New York: William Morrow, 1991.

———. "Industrial Evolution." Interview with Michael Tortorello. *Rain Taxi* 3 (Summer 1998): 34 pars. 8 May 1999 <http://www.raintaxi.com/powers.htm>.

Prigogine, Ilya. *The End of Certainty: Time, Chaos, and the New Laws of Nature.* In collaboration with Isabelle Stengers. New York and London: Free Press, 1997.

———. *From Being to Becoming: Time and Complexity in the Physical Sciences.* San Francisco: W. H. Freeman, 1980.

———. "Order Out of Chaos." Livingston 41–60.

Prigogine, Ilya, and Isabelle Stengers. *Order Out of Chaos: Man's New Dialogue with Nature.* New York: Bantam, 1984.

———. "Postface: Dynamics from Leibniz to Lucretius." Serres, *Hermes* 135–155.

Pynchon, Thomas. *The Crying of Lot 49.* New York: Harper & Row, 1986.

———. *Gravity's Rainbow.* New York: Viking, 1973.

———. "Is It O.K. to Be a Luddite?" *New York Times Book Review* 28 Oct. 1984: 1+.

———. *Mason & Dixon.* New York: Henry Holt, 1997.

———. *Slow Learner.* Boston: Little, Brown, 1984.

———. *Vineland.* Boston: Little, Brown, 1990.

Queneau, Raymond. *Exercises in Style.* Trans. Barbara Wright. New York: New Directions, 1981. Trans. of *Exercices de style.* Paris: Gallimard, 1947.

Redding, Arthur F. "Bruises, Roses: Masochism and the Writing of Kathy Acker." *Contemporary Literature* 35 (1994): 281–304.

Reed, Henry. *Collected Poems.* Ed. Jon Stallworthy. Oxford, Eng.: Oxford University Press, 1991.

Retallack, Joan. "Poethics of a Complex Realism." Perloff and Junkerman 242–73.

Rice, Thomas Jackson. *Joyce, Chaos, and Complexity.* Urbana and Chicago: University of Illinois Press, 1997.

Robbe-Grillet, Alain. *Jealousy.* Trans. Richard Howard. 1957. New York: Grove Press, 1959.

———. *The Voyeur.* Trans. Richard Howard. 1955. New York: Grove Press, 1958.

Rosenberg, Martin E. "Invisibility, the War Machine and Prigogine: Physics, Philosophy and the Threshold of Historical Consciousness in Pynchon's Zone." *Pynchon Notes* 30–31 (Spring-Fall 1992): 91–138.

Safire, William. "Return of The Luddites." *New York Times Magazine* 6 Dec. 1998: 34–35.

Sales, Nancy Jo. "Meet Your Neighbor, Thomas Pynchon." *New York* 11 Nov. 1996: 60–64.

Saltzman, Arthur M. "The Figure in the Static: *White Noise*." *Modern Fiction Studies* 40 (1994): 807–26.

———. "Kathy Acker's Guerrilla Mnemonics." *This Mad "Instead": Governing Metaphors in Contemporary American Fiction*. Columbia: University of South Carolina Press, 2000. 110–28.

———. "Magisters Ludi: *The Universal Baseball Association, Inc.* and *The Music of Chance*." *The Novel in the Balance*. Columbia: University of South Carolina Press, 1993. 60–82.

Schiavetta, Bernardo. "Toward a General Theory of the Constraint." *electronic book review* 10 (1999): 23 pars. 17 June 2000 <http://www.altx.com/ebr/ebr10/10sch.htm>.

Schuber, Stephen P. "Rereading Pynchon: Negative Entropy and 'Entropy.'" *Pynchon Notes* 13 (October 1983): 47–60.

Schwartz, Richard Alan. "Postmodernist Baseball." *Modern Fiction Studies* 33 (1987): 135–49.

Seed, David. "Order in Thomas Pynchon's 'Entropy.'" Bloom, *Thomas Pynchon* 157–74.

Seltzer, Michael. *Terrorist Chic: An Exploration of Violence in the Seventies*. New York: Hawthorn, 1979.

Serres, Michel. "Dream." Trans. Paisley Livingston. Livingston 225–39.

———. *Hermes: Literature, Science, Philosophy*. Ed. Josué V. Harari and David F. Bell. Baltimore: Johns Hopkins University Press, 1982.

———. "Literature and the Exact Sciences." *SubStance* 59 (1989): 3–34.

Shapard, Robert, and James Thomas, eds. *Sudden Fiction: American Short-short Stories*. Salt Lake City, UT: G. M. Smith, Peregrine Smith Books, 1986.

Slusser, George. "Literary MTV." McCaffery, *Storming the Reality Studio*. 334–42.

Snow, C. P. *The Two Cultures and the Scientific Revolution*. New York: Cambridge University Press, 1961.

Sokal, Alan, and Jean Bricmont. *Fashionable Nonsense: Postmodern Intellectuals' Abuse of Science*. New York: Picador, 1998.

Sorrentino, Gilbert. "An Artist Makes Things." Interview with Alexander Laurence. *Arshile* 4 (1995): 95–113.

———. "Genetic Coding." *Something Said: Essays by Gilbert Sorrentino*. San Francisco: North Point Press, 1984. 263–66.

———. *Pack of Lies. Odd Number, Rose Theatre, and Misterioso*. Normal, IL: Dalkey Archive Press, 1997.

Stein, Gertrude. "Composition as Explanation." *Selected Writings of Gertrude Stein*. Ed. Carl Van Vechten. New York: Vintage, 1990. 510–23.

Stephenson, Neal. *Snow Crash*. New York: Bantam, 1992.

Stevens, Wallace. *The Collected Poems of Wallace Stevens*. 1954. New York: Knopf, 1982.

———. *Letters of Wallace Stevens*. Ed. Holly Stevens. New York: Knopf, 1966.

———. *The Necessary Angel*. New York: Vintage, 1951.

Strehle, Susan. *Fiction in the Quantum Universe*. Chapel Hill and London: University of North Carolina Press, 1992.

Tabbi, Joseph. *Postmodern Sublime: Technology and American Writing from Mailer to Cyberpunk*. Ithaca and London: Cornell University Press, 1995.

Tabbi, Joseph, and Michael Wutz, eds. *Reading Matters: Narratives in the New Media Ecology*. Ithaca and London: Cornell University Press, 1997.

Tanner, Tony. "Caries and Cabals." Levine and Leverenz 49–67.

Todd, Olivier. *Albert Camus: A Life*. Trans. Benjamin Ivry. New York: Knopf, 1997.

Tölölyan, Khachig. "War as Background in *Gravity's Rainbow*." Clerc 31–67.

Troika, Lynn Quitman. *Simon and Schuster Handbook for Writers*. 4th ed. Upper Saddle River, NJ: Prentice Hall, 1996.

Unsigned editorial. *Washington Post National Weekly Edition* 13 Jan. 1997: 24.

Waldrop, M. Mitchell. *Complexity: The Emerging Science at the Edge of Order and Chaos*. New York: Simon & Schuster, 1992.

Wallace, Ronald. "The Great American Game: Robert Coover's Baseball." *Essays in Literature* 5 (1978): 103–18.

Watson, Steven. *Strange Bedfellows: The First American Avant-Garde*. New York: Abbeville Press, 1991.

Weisenburger, Stephen. *A "Gravity's Rainbow" Companion: Sources and Contexts for Pynchon's Novel*. Athens and London: University of Georgia Press, 1988.

White, Edmund. "Their Masks, Their Lives—Harry Mathews's *Cigarettes*." *Review of Contemporary Fiction* 7.3 (1987): 77–81.

Wiener, Norbert. *The Human Use of Human Beings: Cybernetics and Society*. 1954. New York: Da Capo Press, 1988.

Williams, William Carlos. *The Collected Poems of William Carlos Williams, Volume II 1939–1962*. Ed. Christopher MacGowan. New York: New Directions, 1988.

———. "George Antheil and the Cantilène Critics." *Selected Essays of William Carlos Williams*. New York: New Directions, 1969. 57–61.

Wilson, E. O. "Back from Chaos." *Atlantic Monthly* Mar. 1998: 41–62.

Wood, James. "Black Noise." Rev. of *Underworld*, by Don DeLillo. *New Republic* 10 Nov. 1997: 38–44.

———. "Levity's Rainbow." Rev. of *Mason & Dixon*, by Thomas Pynchon. *New Republic* 4 August 1997: 32–38.

Woolf, Virginia. "Mr. Bennett and Mrs. Brown." *The Captain's Death Bed and Other Essays*. Ed. Leonard Woolf. London: Hogarth Press, 1950. 90–111.

Yeats, W. B. *The Poems of W. B. Yeats*. Ed. Richard J. Finneran. New York: Macmillan, 1983.

Yorke, James, and Tien-Yien Li. "Period Three Implies Chaos." *American Mathematical Monthly* 82 (1975): 985–92.

Index

About the Author

Joseph Conte is Professor of English at SUNY at Buffalo and author of *Unending Design: The Forms of Postmodern Poetry.*